D0929651

Undergraduate Texts in Mathematics

Springer

New York
Berlin
Heidelberg
Barcelona
Budapest
Hong Kong
London
Milan
Paris
Santa Clara
Singapore
Tokyo

Undergraduate Texts in Mathematics

Anglin: Mathematics: A Concise History and Philosophy.
Readings in Mathematics.

Anglin/Lambek: The Heritage of Thales.
Readings in Mathematics.

Apostol: Introduction to Analytic Number Theory. Second edition.

Armstrong: Basic Topology.

Armstrong: Groups and Symmetry.

Axler: Linear Algebra Done Right.

Beardon: Limits: A New Approach to Real Analysis.

Bak/Newman: Complex Analysis. Second edition.

Banchoff/Wermer: Linear Algebra Through Geometry. Second edition.

Berberian: A First Course in Real Analysis.

Brémaud: An Introduction to Probabilistic Modeling.

Bressoud: Factorization and Primality Testing.

Bressoud: Second Year Calculus.
Readings in Mathematics.

Brickman: Mathematical Introduction to Linear Programming and Game Theory.

Browder: Mathematical Analysis: An Introduction.

Buskes/van Rooij: Topological Spaces: From Distance to Neighborhood.

Cederberg: A Course in Modern Geometries.

Childs: A Concrete Introduction to Higher Algebra. Second edition.

Chung: Elementary Probability Theory with Stochastic Processes. Third edition.

Cox/Little/O'Shea: Ideals, Varieties, and Algorithms. Second edition.

Croom: Basic Concepts of Algebraic Topology.

Curtis: Linear Algebra: An Introductory Approach. Fourth edition.

Devlin: The Joy of Sets: Fundamentals of Contemporary Set Theory. Second edition.

Dixmier: General Topology.

Driver: Why Math?

Ebbinghaus/Flum/Thomas: Mathematical Logic. Second edition.

Edgar: Measure, Topology, and Fractal Geometry.

Elaydi: Introduction to Difference Equations.

Exner: An Accompaniment to Higher Mathematics.

Fine/Rosenberger: The Fundamental Theory of Algebra.

Fischer: Intermediate Real Analysis.

Flanigan/Kazdan: Calculus Two: Linear and Nonlinear Functions. Second edition.

Fleming: Functions of Several Variables. Second edition.

Foulds: Combinatorial Optimization for Undergraduates.

Foulds: Optimization Techniques: An Introduction.

Franklin: Methods of Mathematical Economics.

Gordon: Discrete Probability.

Hairer/Wanner: Analysis by Its History.
Readings in Mathematics.

Halmos: Finite-Dimensional Vector Spaces. Second edition.

Halmos: Naive Set Theory.

Hämmerlin/Hoffmann: Numerical Mathematics.
Readings in Mathematics.

Hijab: Introduction to Calculus and Classical Analysis.

Hilton/Holton/Pedersen: Mathematical Reflections: In a Room with Many Mirrors.

Iooss/Joseph: Elementary Stability and Bifurcation Theory. Second edition.

Isaac: The Pleasures of Probability.
Readings in Mathematics.

(continued after index)

David A. Singer

Geometry: Plane and Fancy

With 117 Figures

Springer

David A. Singer
Department of Mathematics
Case Western Reserve University
Cleveland, OH 44106

Mathematics Subject Classification (1991): 54-01, 54A05

Library of Congress Cataloging-in-Publication Data
Singer, David A.
Geometry : plane and fancy / David A. Singer
p. cm. — (Undergraduate texts in mathematics)
Includes bibliographical references and index.
ISBN 0-387-98306-6 (hardcover : alk. paper)
Geometry. I. Title. II. Series.
QA445.S55 1997
516—dc21 97-26383
CIP

Printed on acid-free paper.

Production managed by Victoria Evarretta; manufacturing supervised by Jacqui Ashri.
Photocomposed copy prepared from the author's LaTeX files.
Printed and bound by Edwards Brothers, Inc., Ann Arbor, MI.
Printed in the United States of America.

9 8 7 6 5 4 3 2 1

ISBN 0-387-98306-6 Springer-Verlag New York Berlin Heidelberg SPIN 10635108

Preface

This book is about geometry. In particular, it is about the idea of curvature and how it affects the assumptions about and principles of geometry. That being said, I should mention that the word "curvature" does not even appear until the end of the fifth chapter of the book. Before then, it is hidden within the idea of the sum of the angles in a triangle.

In the course of the text, we consider the effects of different assumptions about the sum of the angles in a triangle. The main conceptual tool is the tiling, or tessellation, of the plane. Changing our assumptions on triangles leads to vastly different consequences, which can be seen (literally) in the geometric patterns that arise in tilings.

The result of this point of view is a text that goes in atypical directions for a geometry book. In the process of looking at geometric objects, I bring in the algebra of complex (and hypercomplex) numbers, some graph theory, and some topology. Nevertheless, my intent is to keep the book at an elementary level. The readers of this book are assumed to have had a course in Euclidean geometry (including some analytic geometry) and some algebra, all at the high-school level. No calculus or trigonometry is assumed, except that I occasionally refer to sines and cosines. On the other hand, the book touches on topics that even math majors at college may not have seen. This occurs in Chapter 5, so it is possible to skip some or all of this. But I think that would be a mistake. While the ideas in that chapter are advanced, the mathematical techniques are not. For me, that chapter was the main reason for writing this book.

Here is a brief summary of the contents:

Chapter 1 is an introduction to non-Euclidean geometry. Euclid's axiomatic system is based on five postulates, of which four are reasonably intuitive. The fifth postulate, however, is quite another story. In the process of attempting to prove that this postulate follows from the others and is therefore unnecessary, mathematicians discovered many equivalent formulations. The one that is central to this text is the statement due to Gerolamo Saccheri (1733): *The sum of the angles in a triangle is equal to two right angles.* Non-Euclidean geometry begins with the negation of this statement. Throughout the text we will be exploring the consequences of assuming that the sum of the angles in a triangle is always equal to, always less than, or always greater than two right angles. The last section presents a "proof" due to Saccheri that the sum of the angles in a triangle cannot be greater than two right angles, and a "proof" due to Adrien-Marie Legendre that the sum of the angles cannot be less than two right angles.

Chapter 2 proceeds from the assumption that the angle sum is always 180°. We consider the process of tiling the plane with regular polygons. Section 2.1 sets up the machinery of isometries and transformation groups. In Section 2.2 we find all regular and semiregular tilings. A curious unsuccessful attempt to tile the plane with pentagons leads to the construction of self-similar patterns and leads to a digression on fractals. The last section introduces complex numbers as a tool for studying plane geometry.

In Chapter 3 we start instead with the assumption that the angle sum is always less than 180°. This is the underlying postulate for hyperbolic geometry; it can be illustrated by the Poincaré disc. Using this model leads to the hyperbolic tilings, including those used by M.C. Escher in some of his artwork. The simplest description of isometries in this model uses fractional linear (Möbius) transformations. So in Section 3.3 we apply the arithmetic of complex numbers to the geometry of these transformations.

Chapter 4 uses the assumption of angle sums greater than 180°, which is the postulate underlying elliptic geometries. Section 4.1 is a brief look at the complications this assumption causes. We explore the possibility of more than one line connecting two points, relating this to the geometry of the sphere. In the second section the problem of tiling the sphere leads to an introduction to graphs. We derive Euler's formula. The third section consists of the classification of regular and semiregular tilings of the sphere, and the construction of regular and semiregular convex polyhedra. The last section looks at the geometry of the projective plane and includes a description of the Möbius band as the set of lines in the plane.

The fifth chapter, like the Fifth Postulate, is quite a bit more complicated than the first four. It includes topics not found in most (any?) elementary geometry books. Section 1 contains Cauchy's theorem, which states that closed convex polyhedral surfaces are rigid. Although this is an advanced theorem, the proof is elementary and relies only on properties of polygons and Euler's formula.

Section 5.2 generalizes the construction of complex numbers from Chapter 2 to hypercomplex numbers (quaternions). Using the arithmetic of such numbers, we look at the problem of figuring out the effect of two consecutive rotations of the sphere about different axes. Along the way, some of the basic ideas of algebra show up. The third section describes the notion of curvature for polyhedra and includes a proof of the polyhedral Gauss–Bonnet theorem. Again this advanced theorem turns out to rely only on Euler's formula.

Chapter 6 is a brief, nontechnical, discussion of how all of the ideas of the previous chapters can be blended together into a more general notion of geometry. The sum of the angles is used to quantify the curvature of a piece of (two-dimensional) space. A general curved space, either polyhedral or "smooth," is allowed to have curvature that varies from place to place. Straight lines give way to geodesics. We briefly examine the mysterious behavior of shortest paths on polyhedra. A few final words about space–time and general relativity close the chapter.

The contents of this book can be covered in a one-semester course; on the other hand, it would be easy to spend a lot more time on some of the topics than such a schedule would permit. Some sections are very easy to omit: in particular, the discussion of Möbius transformations (Section 3.3) and the discussion of quaternions (Section 5.2) can be dropped to reduce the difficulty level. Section 2.2 on complex numbers is used in those two sections but not elsewhere.

Finally, a note about proofs and mathematical rigor. I have attempted to be precise, not vague, about technical issues, but I have generally avoided the "theorem–proof" style of exposition. Geometry is a fascinating subject, which many people find exciting and beautiful. It is better not to sterilize it by obscuring the main ideas in Euclidean formalism. On the other hand, some rigor is absolutely essential to the subject. Failure to be careful about geometric arguments has led to a lot of nonsense. In Chapter 1, I present the "proof" of the parallel postulate. It has been my practice, in teaching the course for which this book forms the basis, to begin by presenting this proof, preceded by the warning that it is not correct. I believe that the best way to understand the need for proof in mathematics is to see a really good false proof. (This one is a beauty, due to no less a mathematician than Legendre!) After that, I expect my students to be able to convince each other and me of the truth of claims they make. I also expect them to challenge me if they are not convinced about claims I make. This is the ideal environment for mathematical rigor.

Contents

CHAPTER 1 Euclid and Non-Euclid

1.1 The Postulates: What They Are and Why

At the beginning of Euclid's monumental thirteen volume text *The Elements* [18], there is a list of 23 *Definitions*, five *Postulates*, and five *Common Notions*. This book focuses on just one of these, the fifth postulate, commonly known as the "Parallel Postulate." Before we can do that, though, it will be necessary to get some idea of what these definitions, etc., are all about. If you are familiar with Euclid's axioms you may be able to skip this section, which is a brief (and perhaps a bit technical) review.

Euclid attempted to give a completely self-contained theory of geometry. (Actually, very little is known about Euclid himself, so this is really just an inference we can draw from reading *The Elements*. He gave "definitions" of the basic objects of study in geometry. In what follows, I will be quoting directly from the standard source [18]. The first four definitions read:

1. A **point** is that which has no part.
2. A **line** is a breadthless length.
3. The extremities of a line are points.
4. A **straight line** is a line which lies evenly with the points on itself.

These statements are not terribly easy to understand. They are called "definitions," but really a better term might be "undefined terms," since we usually define something by referring to other things we already know about. If geometry is to be a self-contained subject, then we start out at the beginning, not assuming anything. That means we do not have geometric objects we already know about, so we can't define things.

Instead, we begin the study of geometry by assuming that there are things called "points" and "lines," etc., which we will be studying. We may have some prior conception of what they are; for instance, we may describe a point as the smallest thing there is, so that it cannot be further divided into smaller parts. But we don't then try to say exactly what we mean by a "part." We try to keep the list of such undefined terms as short as possible, but there have to be some items on the list. As we will be seeing later, it may be necessary from time to time to revise our understanding of what these words "mean."

The third definition can only be understood if we use the word "line" to refer to line *segments*, which begin at a point A and end at a point B, and "complete" lines, which have no ends or "extremities." A line is not necessarily a straight line, and Definition 4 distinguishes a straight line as a special geometric object. What is meant by the definition is anybody's guess; I will give an interpretation that will be useful in this book. If A is a point on a line ℓ and B is another point on the same line, then we may "slide" ℓ along itself so that point A lands on point B while all the points on the line are moved to points on the same line. One may rightly object that I have not defined the concept of sliding. What I really have in mind is the concept of *isometry*, which I discuss in the next chapter.

Next we come to the Postulates and the Common Notions (usually called Axioms). The Postulates are assumptions made specifically about geometry, which are to be taken as true without proof. The Axioms are assumptions about mathematical truth in general, not specific to geometry. Although Euclid gives the Postulates first, let us examine the Common Notions first. I quote from [18], pp. 154–155:

COMMON NOTIONS

1. Things which are equal to the same thing are also equal to one another.
2. If equals be added to equals, the wholes are equal.
3. If equals be subtracted from equals, the remainders are equal.
4. Things which coincide with one another are equal to one another.
5. The whole is greater than the part.

The fourth item in this list seems to be out of place. Unlike the others, it seems to be specifically about geometry. It is also more mysterious than the others. If two things *coincide*, how can they not be equal? One possible explanation (see [18], pp. 224–231) is that what Euclid was saying was that if two objects *can be made to coincide with one another*, then they are equal (or *congruent*). This idea of making one object coincide with another is called the *method of superposition*; Euclid used this method at times but appears to have disliked it. In Chapter 2 we will elaborate on this idea of comparing objects by moving one to the other. Now let us turn to the Postulates.

POSTULATES

Let the following be postulated:

1. To draw a straight line from any point to any point.
2. To produce a finite straight line continuously in a straight line.
3. To describe a circle with any center and distance.
4. That all right angles are equal to one another.
5. That, if a straight line falling on two straight lines make the interior angles on the same side less than two right angles, the two straight lines, if produced indefinitely, meet on that side on which are the angles less than the two right angles.

The first three postulates are not too bad. The first assures us that we can talk about line segments joining any two points. It is generally understood that Euclid meant by this that there is *exactly one* straight line segment passing through two distinct points, no more and no less. The second postulate appears to mean that a line segment is part of exactly one line, no more and no less. The third postulate tells us that there are circles of any center and radius.

Postulate 4 is a bit mysterious. It seems so reasonable that we ought to be able to prove it. First we must review what we know about right angles. Euclid's tenth definition says:

> When a straight line set up on a straight line makes the adjacent angles equal to one another, each of the equal angles is **right**, and the straight line standing on the other is called a **perpendicular** to that on which it stands.

This should not be interpreted to mean that there actually *are* right angles. In fact, Euclid proves that right angles exist in Proposition 11 (of Book I). The definition tells us how to know if we have a right angle. Postulate 4 says that if you've seen one, you've seen them all. It might seem that the method of superposition could be used to prove this postulate. However, Euclid never stated the principle of superposition, and it can be argued that Postulate 4 is needed to justify the principle. Said another way, the postulate can be interpreted as saying that space is *homogeneous*, that one portion of space looks the same as any other portion of space. We will explore this idea in more detail in Chapter 6.

Another question arises from this discussion. Why should we care whether or not we can prove Postulate 4? Since it is so reasonable, why not just assert that it is true (as Euclid did) and be done with it? To answer this question, we must look at the idea of an axiomatic system. We start with a list of undefined terms and with a list of assumptions about those terms; we call the assumptions axioms or postulates. Then using these assumptions and the laws of logic we develop a body of theorems, each proved by using the axioms and the theorems we have already proved. We need axioms to start up this process; otherwise we would have nothing to use in proving the first theorem.

To be useful, the axioms we choose have to have four important properties:

0. They should be about something. This is not a mathematical criterion, but if the axioms are not about something interesting, then neither will anything else be.
1. They should be *consistent*. In other words, one assumption should not contradict other assumptions. If our axioms were inconsistent, we could use logical arguments to deduce nonsense.
2. They should be *complete*. There must be enough assumptions so that we are able to determine what is true and what is false in our axiomatic system.
3. They should be *independent*. We should not be able to prove one of the axioms from the others.

If we want a system to be consistent, it is best not to make too many axioms. The more axioms, the more possibility that some of them will clash. On the other hand, if we want a system to be complete, then we need to make enough assumptions. So there is a tug of war between consistency and completeness.

If an axiom can be proved from the others, we can throw it out without reducing the scope of our theory. By shortening the list of axioms, we have a better chance of figuring out whether they are consistent. But suppose an axiom cannot be proved from the others. If we delete it as an axiom and try to prove it as a theorem instead, we will not succeed. That means the shortened list of axioms is no longer complete. Therefore, independence of a collection of axioms means that we cannot afford to make the list of axioms any shorter.

Independence is desirable, but not really essential to a system. To assume certain theorems as axioms does not change the theory, and often it makes the theory easier to understand. On occasion, in later sections of this book, I will announce that something is a "fact." What I am doing in effect is adding a theorem to my list of postulates to keep us from getting bogged down in details of proofs. As long as I am actually able back these facts up with proofs, this does no harm.

How can we decide whether an axiomatic system satisfies all of these conditions? TOUGH QUESTION! In practice, it may be difficult or impossible in a particular case to answer this question. We are not going to worry about the consistency or completeness of Euclidean geometry, but we will worry a bit about independence. If we could prove that all right angles are equal, then the fourth postulate would not be independent of the others and we ought to throw it out. Of course, we don't have to throw it out, since it doesn't change our theory. If we could prove that right angles are *not* all equal, then the postulates would be inconsistent.

Which brings us to the fifth postulate. This one takes about three times as many words to state as any of the others, and it is rather mysterious. Euclid appeared to be aware of the special status of the postulate, since he deliberately avoided using it in the proofs of the first 28 propositions.

We can worry about whether the fourth postulate is independent of the first three, but it seems true to us in any event. The trouble with the fifth postulate is that it is not so obviously true. After all, what is it saying? In the diagram, assume that $\angle ABC + \angle BAD < 180°$. Then if we extend the lines through AD and BC far enough, they will meet. But what if the sum of the angles is really, *really* close to 180°? How do we know they will meet?

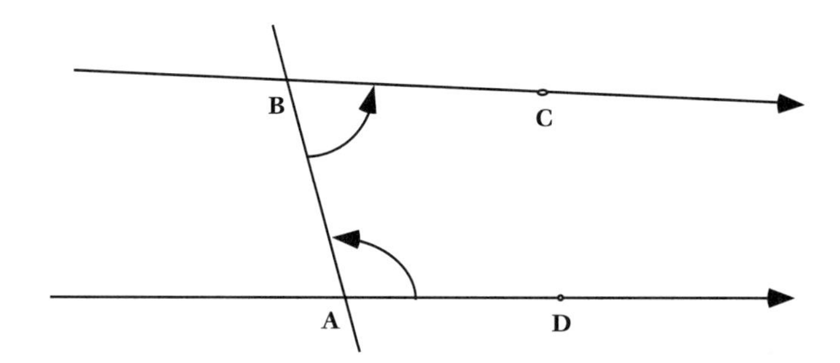

Problem (for readers who know some trigonometry!)
Suppose line segment AB has length 1 in., $\angle BAD = 100°$ *and* $\angle ABC = 79°59'59''$. *How far is it from point A to the point of intersection of the two lines according to Euclidean geometry? Guess the answer before calculating. Do you want to check your answer by construction?*

Euclid's fifth postulate was a source of intense interest and study for over two thousand years. In the next section we will look at some of the ways various mathematicians reformulated the postulate. In the third section, we will see a "proof" of the postulate. However, in the early nineteenth century, several mathematicians independently showed that the fifth postulate is in fact independent of the remaining axioms and postulates.

What this implies is not only that we cannot prove the parallel postulate from the other postulates, but also that we cannot *disprove* it either. Euclid was absolutely right in making it a postulate, since otherwise the question of whether such lines meet or not could not be resolved. It is ironic that a famous book by the eighteenth-century mathematician Girolamo Saccheri, in which the author claimed to demonstrate that the fifth postulate was a logical consequence of the others, was entitled "Euclides ab omni naevo vindicatus" (Euclid vindicated from every flaw). In fact, one could more accurately say that Euclid's vindication came in the next century, when Karl Friedrich Gauss, Nikolai Ivanovich Lobachevsky,[1] and János Bolyai, among others,[2] discovered that one could replace the fifth postulate by its *negation* and still have a consistent theory of geometry. Of course, this new theory is different from the one Euclid proposed; hence the name *Non-Euclidean Geometry*. But as G.B. Halsted wrote in his introduction to the English translation of Bolyai's *The Science of Absolute Space* (this and Lobachevsky's paper are reprinted in [4]):

> In the brilliant new light given by Bolyai and Lobachevski we now see that Euclid understood the crucial character of the question of parallels.
>
> There are now for us no better proofs of the depth and systematic coherence of Euclid's masterpiece than the very things which, their cause unappreciated, seemed the most noticeable blots on his work.

Originally, there was only one form of non-Euclidean geometry, now commonly known as *hyperbolic geometry*. Lobachevsky published his paper "Geometrical Researches on the Theory of Parallels" in 1829. About two years later, Bolyai's work appeared as an appendix to a book by his father, Wolfgang. The geometry of Lobachevsky and Bolyai is based on the assumption that through a point not on a given line it is possible to find more than one line not meeting the given line. We will explore this assumption in Chapter 3.

It was not until 1854 that a different geometry appeared. Bernhard Riemann described in his dissertation a geometry based on the idea that any two lines must intersect. This geometry is more complicated than hyperbolic geometry; it is actually necessary to modify Euclid's first two postulates as well as the fifth postulate. In fact, there are really two different geometries, sometimes called *single elliptic geometry* and *double elliptic geometry*. We will explore these geometries in Chapter 4.

---

1. There are numerous English spellings of this name.
2. E.g., Ferdinand Karl Schweikart. See [4].

Many people have proposed alternative sets of axioms to replace those of Euclid. Of these, the most famous is due to David Hilbert. They are more precise than Euclid's five postulates, but they are also quite a bit more complex. Here are Hilbert's axioms of plane geometry, taken from [21], pp. 3–26 (I omit the axioms of solid geometry):

I. Axioms of Incidence

I, 1. For every two points A, B there exists a line that contains each of the points A, B.

I, 2. For every two points A, B there exists no more than one line that contains each of the points A, B.

I, 3. There exist at least two points on a line. There exist at least three points that do not lie on a line.

These axioms say that two points determine a line and that not all points are on the same line.

II. Axioms of Order

II, 1. If a point B lies between a point A and a point C then the points A, B, C are three distinct points of a line, and B then also lies between C and A.

II, 2. For two points A and C, there always exists at least one point B on the line AC such that C lies between A and B.

II, 3. Of any three points on a line there exists no more than one that lies between the other two.

II, 4. Let A, B, C be three points that do not lie on a line and let a be a line in the plane ABC which does not meet any of the points A, B, C. If the line a passes through a point of the segment AB, it also passes through a point of the segment AC, or through a point of the segment BC. Expressed intuitively, if a line enters the interior of a triangle, it also leaves it.

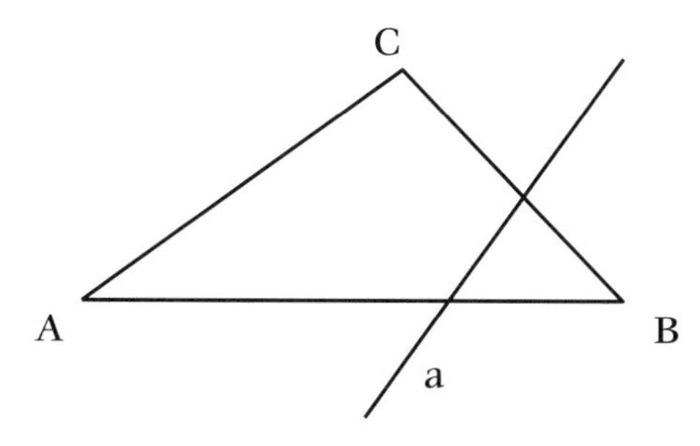

These axioms establish the idea of a point being "between" two other points (II,1 and II,3) and that a line does not have any ends (II,2).

Axiom II,4 is known as *Pasch's Postulate.* Moritz Pasch was the first mathematician to identify assumptions that Euclid made implicitly but did not state explicitly. Euclid assumed in the proof of his Proposition 21 (of Book 1) that a line entering a triangle at a vertex must, if extended sufficiently, intersect the opposite side.

A subtle consequence of these axioms is that between any two points on a line there must be (infinitely many) other points. Perhaps you would like to prove this for yourself.

Problem

Use Axioms I,3, II,2, II,3, and II,4 to show that for any two points A and C there is a point B between them on the line segment AC. [HINT: Start with A and C and use I,3 and II,2 to find other points. Eventually, use II,4 to find an intermediate point on the segment AC.]

III. Axioms of Congruence

III, 1. If A, B are two points on a line a, and A' is a point on the same or on another line a', then it is always possible to find a point B' on a given side of the line a' through A' such that the segment AB is congruent or equal to the segment $A'B'$. In symbols $AB \equiv A'B'$.

III, 2. If a segment $A'B'$ and a segment $A''B''$ are congruent to the same segment AB, then the segment $A'B'$ is also congruent to the segment $A''B''$, or briefly, if two segments are congruent to a third one they are congruent to each other.

III, 3. On the line a let AB and BC be two segments which except for B have no point in common. Furthermore, on the same or another line a' let $A'B'$ and $B'C'$ be two segments which except for B' also have no point in common. In that case, if $AB \equiv A'B'$ and $BC \equiv B'C'$ then $AC \equiv A'C'$.

III, 4. Let $\angle(h, k)$ be an angle in the plane α and a' a line in the plane α and let a definite side of a' in α be given. Let h' be a ray on the line a' that emanates from the point O'. Then there exists in the plane α one and only one ray k' such that the angle $\angle(h, k)$ is congruent or equal to the angle $\angle(h', k')$ and at the same time all interior points of the angle $\angle(h', k')$ lie on the given side of a'. Symbolically $\angle(h, k) \equiv \angle(h', k')$. Every angle is congruent to itself, i.e., $\angle(h, k) \equiv \angle(h, k)$.

III, 5. If for two triangles ABC and $A'B'C'$ the congruences

$$AB \equiv A'B', AC \equiv A'C', \angle BAC \equiv \angle B'A'C'$$

hold, then the congruence

$$\angle ABC \equiv \angle A'B'C'$$

is also satisfied.

Axiom III,1 can be used to show the existence of circles. Axiom III,5 can be used to prove the SAS congruence law. Axiom III,4 is difficult to read. It says that given a ray h' emanating from a point O' there are two other rays that make a specified angle with h'. To distinguish between the two angles, think of h' as being part of a line a'. Then one of the angles lies on each side of this line.

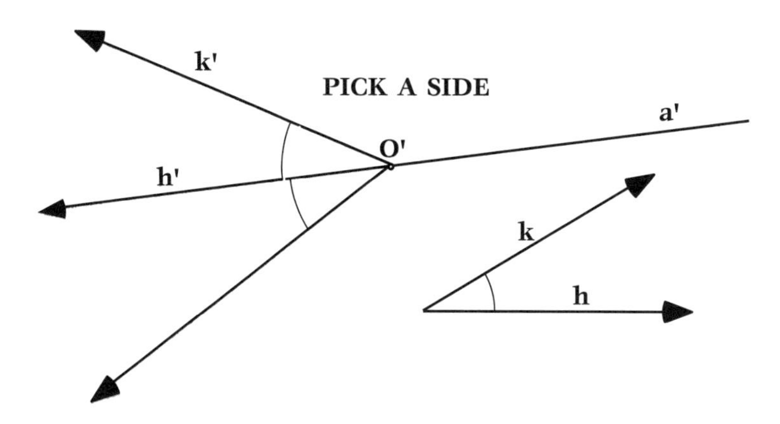

IV. Euclid's Axiom

Let a be any line and A a point not on it. Then there is at most one line in the plane, determined by a and A, that passes through A and does not intersect a.

V. Axioms of Continuity

V, 1 (Archimedes' Axiom) If AB and CD are any segments then there exists a number n such that n segments CD constructed contiguously from A, along the ray from A through B, will pass beyond the point B.

V, 2 (Completeness). An extension of a set of points on a line with its order and congruence relations that would preserve the relations existing among the original elements as well as the fundamental properties of line order and congruence that follows from Axioms I-III, and from V, 1 is impossible.

We saw in the problem above that between any two points on a line there has to be another point. The real number line has this property, but so does the rational number line. If we choose a unit of measurement for lengths, then it is possible to show that not all lengths are rational. (This fact was known to Pythagoras). However, it requires an extra axiom to show that all real numbers are possible lengths. This is the completeness axiom. (An example of a geometry in which this axiom fails is described later in this section.)

Problem

Prove Euclid's fourth postulate from Hilbert's axioms. [HINT: Use SAS to prove that if two angles are equal, then their supplements must be equal. That proves that straight angles are equal.]

Models. Hilbert did not merely formulate a system of axioms for geometry; he also addressed the questions of consistency and independence of the axioms. How can one show that axioms are consistent? Hilbert's approach was through the use of a *model*. There are certain undefined terms, including *point, line, on, between,* and *congruent*. Hilbert gave an interpretation of each of these using arithmetic: he used *Cartesian geometry*.

Suppose we represent a **point** by a pair of real numbers (a, b). It takes two numbers to describe the location of a point. To describe a line, we are going to need three numbers. (Think about why this is so!) The ratio $(u : v : w)$ of three numbers defines a **line**, provided that u and v are not both zero. (By a ratio, we mean that $(u : v : w)$ and $(u' : v' : w')$ should be thought of as the same if the ratios of corresponding pairs of numbers are the same.) A point (x, y) is **on** a line $(u : v : w)$ if the relation $ux + by + w = 0$ holds.

This is, of course, what we do in analytic geometry. It is now a reasonably easy process to check that all of the postulates of Hilbert are true if we interpret them as statements in analytic geometry. We say that analytic geometry is a *model* for Hilbert's geometry.

Now suppose the axioms of Hilbert were not consistent. Then it would also have to be true that the laws of ordinary arithmetic are inconsistent, since each of the Hilbert axioms can be deduced from the laws of arithmetic. So as long as we assume that arithmetic has no inconsistencies, then neither does Euclidean geometry.

Models can also be used to prove the independence of axioms. For example, Hilbert described ([21], pp. 29–30) a model for geometry in which every axiom holds except the last one (completeness). The model he used is just like ordinary analytic geometry, except that he only allows numbers that can be constructed from whole numbers using addition, subtraction, multiplication, division, and the operation $\sqrt{1+x^2}$, where x is any number already produced. So we can get the number $\sqrt{13}$ because it equals $2\sqrt{1+(3/2)^2}$. But we cannot get *all* real numbers. For example, we cannot get the number π, because π is not an algebraic number (a root of a polynomial with rational coefficients.)

Hilbert showed that his model satisfies all of the axioms of geometry except the Completeness Axiom, which is false in this model. If we take as our axioms of geometry everything except V,2 plus the *negation* of V,2, we get something that is as consistent as arithmetic. This proves that the Completeness Axiom is independent of the other axioms. Why is that? Because if we could prove the Completeness Axiom, then we would find an inconsistency in the model of geometry just described, and that would mean that arithmetic is inconsistent.

In Chapter 3 we will see the Poincaré model of the hyperbolic plane, which is based on Euclidean geometry. This model shows that if hyperbolic geometry is inconsistent, then so is Euclidean geometry (and hence, arithmetic).

1.2 The Parallel Postulate and its Descendants

In this section we will look at alternative formulations of the parallel postulate. Originally, many of these formulations were tacit assumptions made by mathematicians attempting to prove the parallel postulate. Some of them will shed light on the role the fifth postulate plays in describing the geometry of space. References for this section include [4], chapters I and II; [18], 202–220; and [36], chapter II. All of the statements are true in Euclidean geometry, so the key to understanding them is to figure out how each assumption implies the parallel postulate. In the more interesting cases, this is not altogether obvious!

The first example is the formulation used by Hilbert as Axiom IV (see the previous section). It states:

Hypothesis 1 *Through a given point can be drawn only one parallel to a given line.*

The language of this statement is tricky. What does the word "parallel" mean? According to Euclid:

Parallel straight lines are straight lines which, being in the same plane and being produced indefinitely in both directions, do not meet one another in either direction.

This is the definition we will be using. Other definitions sometimes beg the question. For instance, according to Proclus, in his *Commentary on the First Book of Euclid*, Posidonius defined parallel lines to be "those which, (being) in one plane, neither converge nor diverge, but have all the perpendiculars equal which are drawn from the points of one line to the other"([4], p. 190). This is all well and good, but then we need to know whether lines that do not meet are parallel, and also whether parallel lines actually exist!

Hypothesis 2 *There exists a pair of straight lines everywhere equally distant from one another.*

Proclus gave a proof of the parallel postulate based on an apparently milder hypothesis than (2). Suppose two lines a and b have a common perpendicular AB. If we take any other point C on line b and drop a perpendicular from it to line a, meeting the line at point D, then Hypothesis 2 would claim that AB and CD are congruent and that CD is a common perpendicular of the two lines. Suppose we assume instead only that the length of CD cannot be arbitrarily large. In other words, if AB is one inch long, then any segment CD from a point on b dropped perpendicularly to D can be no larger than, say, five trillion miles. Then Proclus argues that any line through C must meet line a.

Suppose CG is such a line. As the lines CF and CG are extended indefinitely, the distance between points on one and points on the other increases without bound. So eventually the distance from point G to the line b will be greater than the distance from line a to line b. So the line CG will cut line a.

This argument is basically correct, but it relies on:

Hypothesis 3 *If two lines a and b are parallel, then the distance from a point on one line to the other cannot be made arbitrarily large.*

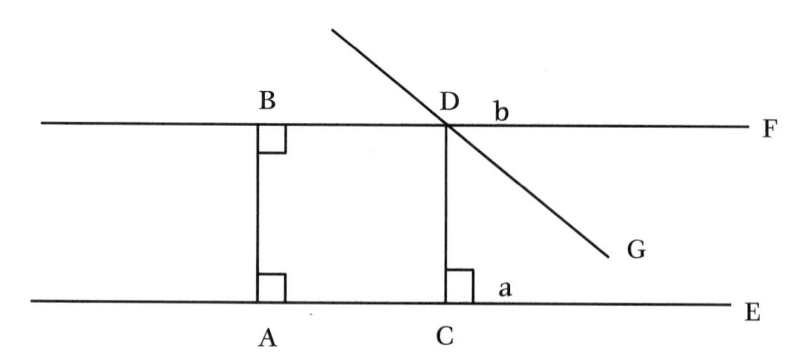

Hypothesis (2) obrviously implies Hypothesis (3), which by Proclus' argument implies Hypothesis (1). So we can conclude that all three of these statements are equivalent to the parallel postulate.

Actually, even the following assumption is enough to prove the parallel postulate:

Hypothesis 4 *If a and b are parallel lines in the plane, then there is a segment AB from a point A on a to a point B on b which is a common perpendicular.*

Problem
Prove that Hypothesis 4 implies the parallel postulate. Hint: It will probably help to read on a bit before trying this one!

The following hypothesis is apparently due to the thirteenth-century Persian mathematician Nasiraddin at-Tusi ([18], pp. 208–209). It is a central idea in this book.

Hypothesis 5 *In any triangle the three angles are together equal to two right angles.*

Using this hypothesis, it is possible to prove Euclid's fifth postulate. Since we will be using this fact about triangles throughout the book, we really should see a proof.

Girolamo Saccheri later noted that it is actually enough to find *one* triangle whose angle sum is equal to two right angles in order to prove the parallel postulate. (See the next section.)

Problem
Assume Hypothesis 5. Let a and b be parallel lines with a common perpendicular AB. Drop a perpendicular from some point C on b to a, meeting a at point D. Conclude that CD is also perpendicular to line b, and that CD is congruent to AB.

Theorem 1.2.1 *If the sum of the angles in a triangle is always equal to* $180°$ *(two right angles), then the parallel postulate holds.*

Proof
Let ℓ be a line and P a point not on the line. Our goal is to show that exactly one line through P fails to intersect ℓ.

Drop a perpendicular from P to the line, meeting the line at point A_1. Construct the line PQ through P perpendicular to PA_1. This line cannot meet ℓ, so we must show that no other line through P fails to meet ℓ.

Construct segment A_1A_2 along ℓ congruent to PA_1. (This can be accomplished by drawing the circle with center at A_1 passing through P.) Then angle $\angle A_1PA_2 = \theta = 45°$. So $\angle QPA_2$ is also $45°$. Next, construct A_2A_3 congruent to PA_2. The triangle PA_2A_3 is isosceles, so the two base angles

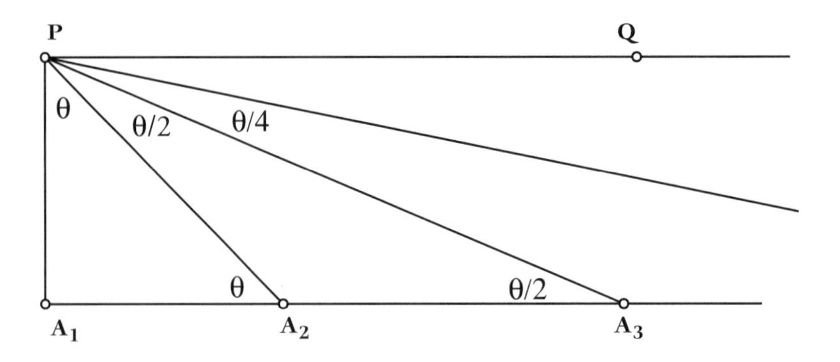

must each be $\frac{\theta}{2}$. (Of course, we are assuming that the sum of the angles in a triangle is 180°.) This means that the line PA_3 bisects the angle $\angle QPA_2$.

We can repeat this construction over and over again, getting a sequence of line segments PA_n making smaller and smaller angles with line PQ. If PB is any line through P making some angle ϕ with PQ, then the line PA_n can be constructed making an angle smaller than ϕ. That means that the line PB enters the triangle PA_1A_n, and it must intersect the line ℓ. ■

This is an argument worth thinking about for a while. If the sum of the angles in a triangle were always less than 180°, we could still construct the points A_n in exactly the same way, but the angles of the isosceles triangles would be smaller than expected. Thought about another way, as a point A slides along ℓ farther and farther away from A_1, the angle QPA would not shrink down to 0. Instead, the line PA would approach a limiting position PZ.

By symmetry, there should be a pair of lines through P, making equal angles with PQ, not meeting the line AA_1. These two lines should have the property that any line through P making a larger angle with PQ would hit

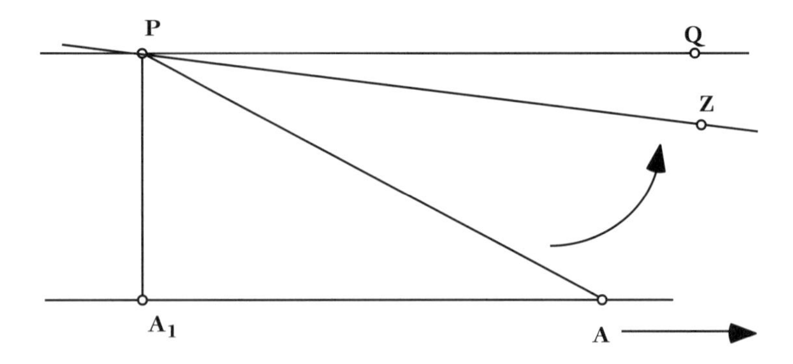

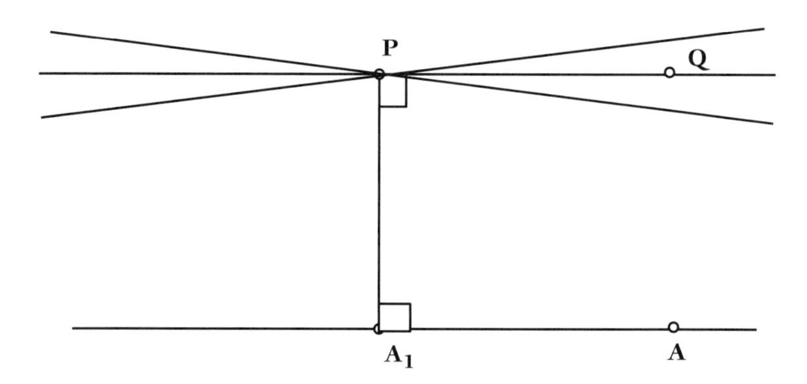

line AA_1 somewhere. This is the basic picture of parallelism in hyperbolic geometry.

There are many other interesting assumptions that turn out to be equivalent to the parallel postulate. Here are a few:

Hypothesis 6 (Wallis) *There exist a pair of similar triangles which are not congruent.*

Hypothesis 7 (Legendre, W. Bolyai) *Given any three points not on a straight line, there exists a circle passing through them.*

Hypothesis 8 (Gauss) *There exist triangles of arbitrarily large area.*

Problem
Find a proof of the parallel postulate from Hypothesis 6.

In the next section, we will use another hypothesis that is equivalent to the parallel postulate. Watch for it!

Problem
What is wrong with the following argument for proving Hypothesis 6? This "proof" is due to Bernhard Friedrich Thibaut (1809). (See [36], pp. 40–41).

Let ABC be a triangle with angles α, β, and γ, and consider the line containing A and B. Rotating the line around the vertex B until it coincides with the line through B and C rotates it through the angle $180° - \beta$. Rotating the line about the vertex C until it coincides with the line through C and A rotates the line through the angle $180° - \gamma$. Finally, rotating the line about the vertex A back to its original position rotates it through a further angle $180° - \alpha$. In all, the line rotates through $540° - \alpha - \beta - \gamma$. Since the line has rotated $360°$, the sum of the interior angles must be $180°$.

1.3 Proving the Parallel Postulate

In order to understand this section, it is vital to keep in mind an important fact: *The "proof" of the parallel postulate presented here is not correct.* In fact, there can be no proof of the parallel postulate that relies only on the other axioms and postulates of Euclid. By the end of this book, it should be clear where the mistake(s) is (are) in this section.

This is not to say that the proof is all wrong. In fact, most of it is quite correct. This is not so surprising, since the author of the proof was the great eighteenth-century mathematician Adrien-Marie Legendre. Legendre's studies in the calculus of variations, elliptic functions, and number theory, among other works, made him one of the leading mathematicians of his day and one whose work is still important today. His textbook *Eléments de Géométrie*, 1794, was widely used in Europe and the United States as a substitute for Euclid. Legendre attempted to prove the parallel postulate in early editions of the text, but he was not satisfied with his own arguments. His final attempt at a proof appeared in 1833 in a paper entitled *Refléxions sur différentes manières de démontrer la théorie des parallèles ou le théorème sur la somme des trois angles du triangle* [Mém. Ac. Sc., Paris, T.XIII, 1833]. It is this proof that we will be examining here. For extensive discussion of proofs of the parallel postulate, see [4], [10], or the briefer treatment in [9].

Earlier in the eighteenth century, Gerolamo Saccheri attempted to use logical reasoning to establish the truth of the parallel postulate. His basic approach was to assume the hypothesis that the postulate is false, and then to show that a contradiction is reached. This method, known as *reductio ad absurdum*, is a very important and powerful one, and in fact, Euclid used it as early as the proof of Proposition 6 of *Elements* ([4], pp. 136, 256). We consider a list of cases, one of which has to be true because all possibilities have been exhausted in the list. We then rule out all but one item on the list. That last survivor must be the true statement.

Saccheri considered the sum of the angles in a triangle. We have already noticed in the last section that the parallel postulate is equivalent to the assumption that this sum must always be 180°. Saccheri's observation is this: There are three possibilities for the sum of the angles in a triangle:

1. The angles in a triangle always sum to 180°.
2. There is some triangle whose angles sum to more than 180°.
3. The sum of the angles in a triangle is always no more than 180°, but there is some triangle whose angles sum to less than 180°.

If we can rule out (2) and (3), then (1) must be true, and the parallel postulate is proved. Legendre's proof is then divided into two pieces. He first rules out the possibility of an angle sum being greater than 180°. This is Saccheri's "Hypothesis of the Obtuse Angle." Actually, Saccheri worked with four-sided polygons with three right angles, but the idea is basically the same. Then Legendre assumes (3) and shows that that leads to an impossibility too. This completes the proof.

Before we begin, we need one very important geometry lemma, which Euclid gives as Proposition 24. Since Euclid does not rely on the parallel postulate until he gets to Proposition 27, this is O.K. to use. We have to be careful never to assume what we are trying to prove! (That, after all, is Legendre's fatal mistake!) Besides needing this lemma now, we will be relying on it in Chapter 5. I like to call this the "caliper lemma;" it says that as you open the caliper, so that the angle is bigger, the ends of the caliper move apart.

Lemma 1.3.1
Suppose triangles ABC and DEF are given with side AB congruent to side DE, side AC congruent to side DF, and $\angle BAC$ smaller than $\angle EDF$. Then side BC is smaller than side EF.

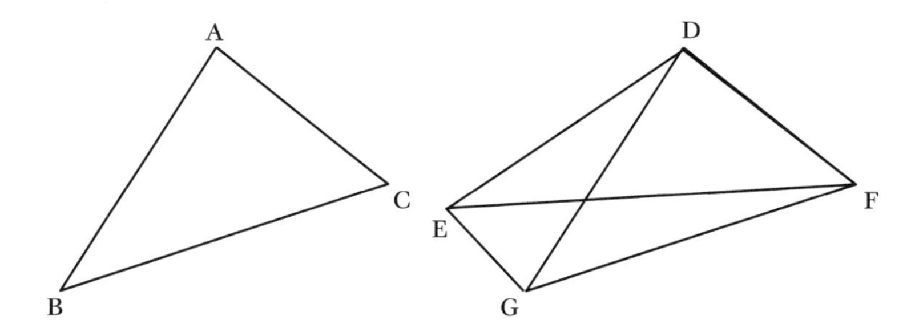

Proof
Construct triangle DGF congruent to triangle ABC. Now, our goal is to show that GF is smaller than EF. The triangle EDG is isosceles, so $\angle DEG = \angle DGE$. Since $\angle DGE$ is smaller than $\angle FGE$, and $\angle DEG$ is larger than $\angle FEG$, it follows that $\angle FEG$ is smaller than $\angle FGE$. In any triangle, the larger side is opposite the larger angle. So side GF is shorter than side EF. ∎

The picture makes the argument easier to follow. But is it an accurate picture? Actually, there is a hidden assumption: Side DG is longer than side DF. Otherwise, the point G might land inside the triangle DEF or even on the line EF.

Problem
Fix the proof of the Lemma.

Theorem 1.3.2 (Saccheri, 1733; proof by Legendre) *The sum of the angles in a triangle is always less than or equal to* $180°$.

Proof
Suppose triangle ABC has angles α, β, and γ with $\alpha+\beta+\gamma > 180°$. If the length of side AB is c, mark off consecutive segments A_1A_2, A_2A_3, A_3A_4, etc. of length c along a fixed line ℓ and construct triangles $A_1C_1A_2$, $A_2C_2A_3$, etc. congruent to ABC with their upper vertices C_i on the same side of the line. Connect each pair of consecutive upper vertices C_iC_{i+1} by a line segment. These line segments all have the same length d.

Now let $\delta = \angle C_1A_2C_2$ between the sides of adjacent triangles. Then $\alpha+\delta+\beta = 180°$, since the angles fill out a straight angle. Now, that tells us that δ must be smaller than γ, so the caliper lemma says that d is smaller than c.

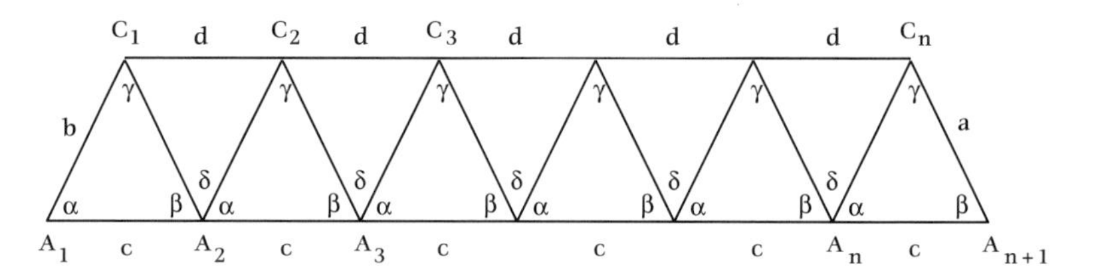

The straight-line distance from A_1 to A_{n+1} is nc. If instead we go from A_1 to C_1, then successively visit $C_2, C_3, \ldots, C_n$, and finally go to A_{n+1}, this path will have length $a+(n-1)d+b$. Since the straight–line path is the shortest path (this follows from Euclid, Proposition 20, [18]), we get the interesting inequality

$$a+b-d > n(c-d).$$

The left–hand side is a positive number (again by Proposition 20), and the quantity $(c-d)$ is also positive. The strange thing is, while both of those quantities are fixed, the number n is not! The more triangles we build, the bigger we make n. But if we make n large enough, we can make the inequality fail. (Technically, we are using the *Archimedean axiom* here. It says that there must be an integer larger than the number $\frac{a+d-c}{c-d}$.) This contradiction proves the theorem. ■

We may assume from now on that the sum of the angles in a triangle is no larger than 180°. If ABC is any triangle, with angles α, β, and γ, let's call the number $180° - \alpha - \beta - \gamma$ the *defect* of the triangle and denote it by the symbol $\delta(ABC)$. This is a number between 0 and 180 (ignoring the units). To finish the proof of the parallel postulate, Legendre wanted to prove that the defect of any triangle has to be 0. Let's examine defects for a moment.

Start with a triangle ABC with angles α, β, and γ. Let D be any point on side BC. The line segment AD cuts the angle α into two pieces, α_1 and

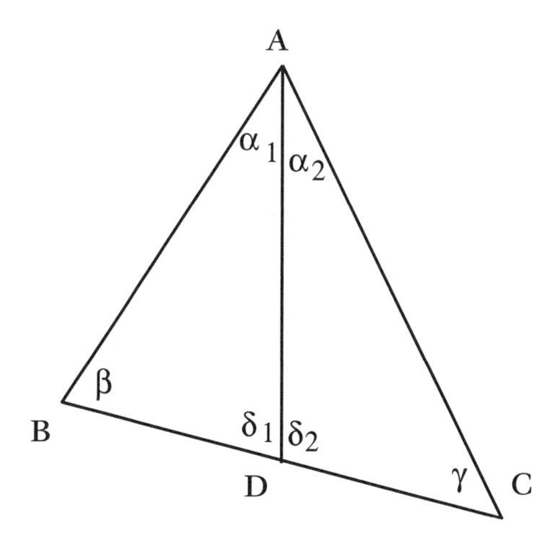

α_2. Denote $\angle ADB$ by δ_1 and denote $\angle ADC$ by δ_2. Then $\delta_1 + \delta_2 = 180$. The defects of the triangles are given by:

$$\begin{aligned}
\delta(ADB) &= 180 - (\delta_1 + \beta + \alpha_1);\\
\delta(ADC) &= 180 - (\delta_2 + \gamma + \alpha_2);\\
\delta(ABC) &= 180 - (\gamma + \beta + \alpha)\\
&= 360 - (\delta_1 + \delta_2) - (\gamma + \beta + \alpha_1 + \alpha_2)\\
&= \delta(ADB) + \delta(ADC).
\end{aligned}$$

This is an example of a general principle: *If a triangle is subdivided into smaller triangles, then the defect of the triangle is the sum of the defects of the smaller triangles.* Using this, Saccheri was able to prove the following:

Lemma 1.3.3
If some triangle has positive defect, then every triangle has positive defect.

Problem
Prove Lemma 1.3.2. One way to do this: Assume that there is a triangle with positive defect. Then show that there is a right triangle with positive defect. Then show that any *right triangle has positive defect. Finally, show that any triangle has positive defect.*

Theorem 1.3.4 (Legendre, 1833) *The sum of the angles in every triangle is exactly* 180°.

Proof
Suppose we have a triangle ABC with defect $\delta(ABC) = \delta > 0$. Reflect ABC across side BC and let D be the image of A. Extend the rays from A to B and C and construct line EF through D meeting the ray AB at E and the ray

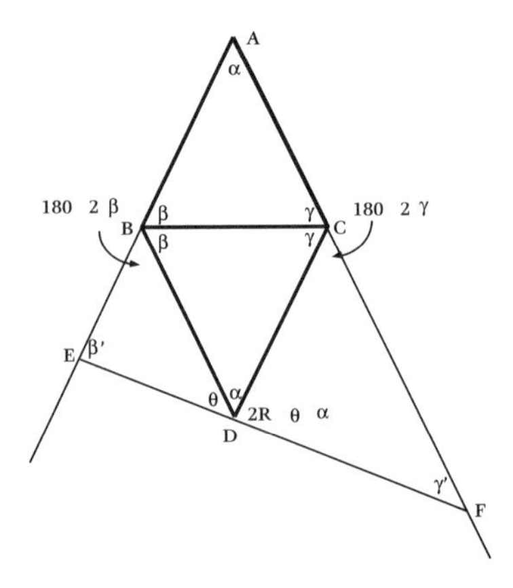

AC at F. The triangle AEF is subdivided into four smaller triangles, two of which are known to have defect δ. It follows that the defect $\delta(AEF) > 2\delta$. Repeating this construction, we may produce a triangle with defect $2^n\delta$ for any positive n. But if n is chosen large enough, then $2^n\delta > 180$. This is impossible! Therefore, there can be no triangle with defect $\delta > 0$. ■

Notwithstanding the theorem just "proved," there are still three possibilities for the sum of the angles in a triangle. In Chapter 2 we will assume that the sum of the angles is always 180°. In Chapter 3 we will assume instead that the sum is always less than 180°. In Chapter 4 we will explore the third possibility, that the sum is always greater than 180°. Chapter 6 will look at a more general idea of geometry in which some triangles can have angle sum larger than 180°, some less than 180°, and some equal to 180°. This will necessitate changing some basic assumptions made in the rest of the book.

Problem

What hidden assumptions were used in proving Theorems 1.3.2 and 1.3.4? Do these assumptions depend on the parallel postulate? If so, how?

Tiling the Plane with Regular Polygons

2.1 Isometries and Transformation Groups

A good starting point for the study of the geometry of the (Euclidean) plane is the concept of *isometry*. To many modern geometers this is the natural way to talk about things like congruence, and we might expect Euclid to have used isometries in his *Elements*. Strangely enough, Euclid fails to mention isometries; yet he appears to use them from the very outset. Proposition 4 ([18], p. 247) states:

If two triangles have the two sides equal to two sides respectively, and have the angles contained by the equal straight lines equal, they will also have the base equal to the base, the triangle will be equal to the triangle, and the remaining angles will be equal to the remaining angles respectively, namely those that the equal sides subtend.

Part way into the proof comes the peculiar argument, "For, if the triangle ABC be applied to the triangle DEF, and if the point A be placed on the point D and the straight line AB on DE, then the point B will also coincide with E, because AB is equal to DE."

What exactly does the word "applied" mean? This question engaged the attention of many mathematicians, who perceived a certain vagueness in the proof of Proposition 4 and proposed clarifications. (For a full discussion, see [18], pp. 224–231 and 249–250.) We will interpret "applied" to mean that there is an isometry carrying one triangle to the other. We will also use the word "congruent" in place of the word "equal," and say that two figures are congruent if one can be applied to the other (by an isometry).

An *isometry* is a correspondence $\mathbf{T}$ that assigns to each point A in the plane a point $A' = \mathbf{T}(A)$ in such a way that for any points A and B, the corresponding line segments AB and $A'B'$ are congruent. We might visualize this by taking a sheet of paper and marking points $A, B, C, \ldots$ on it. Now let's put a transparency over the paper and mark points $A, B, C, \ldots$ on the transparency over the points with the same names. If we now shift the transparency to a new location over the paper, the marked points will lie over new points $A', B', C', \ldots$ on the paper. Since the transparency does not stretch, this is an isometry.

We need the axiom of Pasch [27] here. This says that if A, B, and C are points that do not lie on one line and if A' and B' are two points chosen such that AB and $A'B'$ are congruent, then there exactly two choices of a point C' such that the triangle ABC is congruent to triangle $A'B'C'$.

Two? Well, pick A' and B' on the paper such that A and B on the transparency can be simultaneously superimposed on them. Look where C goes on the paper. That's one. Now *flip the transparency over* and line up A and B with A' and B'. The point labeled C now lies over the second possible location for C'.

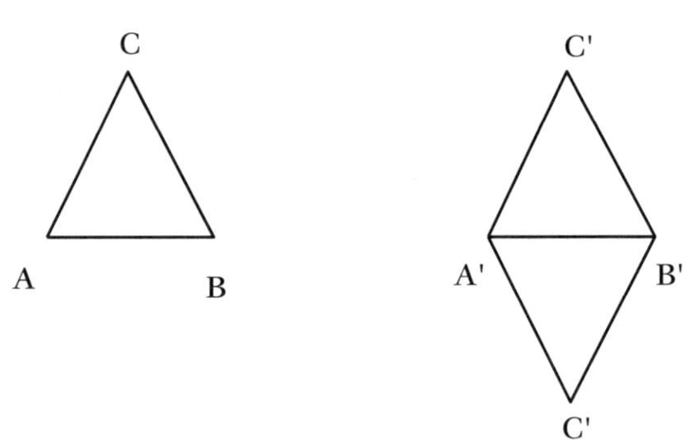

Problem

Suppose we have four points A, B, C, D not all lying on one line. Suppose A' and B' are chosen so that AB and $A'B'$ are congruent. Choose C' such that AC is congruent to $A'C'$ and BC is congruent to $B'C'$. How many choices are there for a point D' such that AD is congruent to $A'D'$, BD is congruent to $B'D'$, and CD is congruent to $C'D'$? Why?

Examples of isometries.

1. A *translation* $\mathbf{T}$ along a line ℓ moves the points on the plane so that each point on ℓ moves the same distance along ℓ to another point

on ℓ. Where do points not on the line move? If B is a point not on ℓ and AB is the perpendicular from B to ℓ, then $A, B, \mathbf{T}(B)$, and $\mathbf{T}(A)$ form a rectangle. The line through B and $\mathbf{T}(B)$ is also carried to itself by the translation. (See the figure below.)

2. A *rotation* $\mathbf{R}$ about a point O through angle θ fixes the point O (that is, $R(O) = O$) and takes each line through O to the line that makes the angle θ with it. If A is any point, then $\Delta AO\mathbf{R}(A)$ is an isosceles triangle with angle θ at the vertex O.
3. A *flip* (or *reflection*) $\mathbf{F}$ across a line ℓ fixes every point on ℓ while moving all points to their mirror images on the other side of ℓ.
4. A *glide reflection* $\mathbf{G}$ is a translation along ℓ followed by a flip across ℓ.

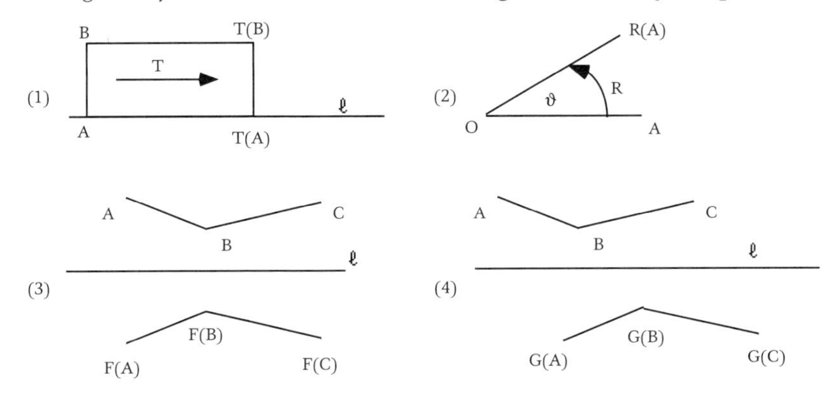

It is a basic fact of Euclidean geometry that these are the only possible isometries of the plane.

An important property of an isometry $\mathbf{U}$ is that it is a *bijection*. This means that for every point A there is exactly one point A_1 that is taken to A by $\mathbf{U}$. It is easy to see that two different points A_1 and A_2 could not be taken to the same point A. After all, if $\mathbf{U}(A_1) = \mathbf{U}(A_2)$, then the line segment from A_1 to A_2 would have to be congruent to the "line segment" from A to itself. It is less easy to see why there has to be *some* point A_1 for which $\mathbf{U}(A_1) = A$.

Problem

Show that the point A_1 can indeed be found. [HINT: Find two points B_1 and C_1 for which $\mathbf{U}B_1$, $\mathbf{U}C_1$, and A form a triangle. Use the last problem.]

If $\mathbf{U}$ is an isometry and $\mathbf{V}$ is another isometry, then we can make a new isometry $\mathbf{VU}$ by the rule

$$\mathbf{VU}(A) = \mathbf{V}(\mathbf{U}(A)).$$

In other words, if $\mathbf{U}$ moves the point A to A' and if $\mathbf{V}$ moves the point A' to A'', then $\mathbf{VU}$ moves A to A''. Notice that what $\mathbf{V}$ does to the point A does not matter in figuring out what $\mathbf{VU}$ does to the point A.

How do we know that **VU** is an isometry? If **U** carries B to B' and **V** carries B' to B'', then **VU** takes B to B''. Since **U** is an isometry, the line segment AB is congruent to $A'B'$. Since **V** is also an isometry, $A'B'$ is congruent to $A''B''$. **VU** carries AB to the congruent segment $A''B''$, which is what an isometry is supposed to do.

If we instead form the isometry **UV**, made by putting the two isometries together in the opposite order, we may not get the same isometry. Here is an example to help us visualize this. Take a triangle ABC with side AB horizontal and point C above it somewhere. Think of the triangle as an arrow pointing up. Suppose **F** is the flip across the (horizontal) line ℓ through A and B, while **R** is the rotation through 90° (counterclockwise) around A. Neither of these moves the point A at all, so the isometries **RF** and **FR** also keep A fixed. But **RF** moves the triangle ABC to an arrow pointing right, while **FR** moves it to an arrow pointing left. (See the figure below.) (This is a bit confusing. To do **FR**, you must first rotate and then flip across the original horizontal line. The rotation has moved ℓ to a new line ℓ', but the flip is still taken across ℓ.)

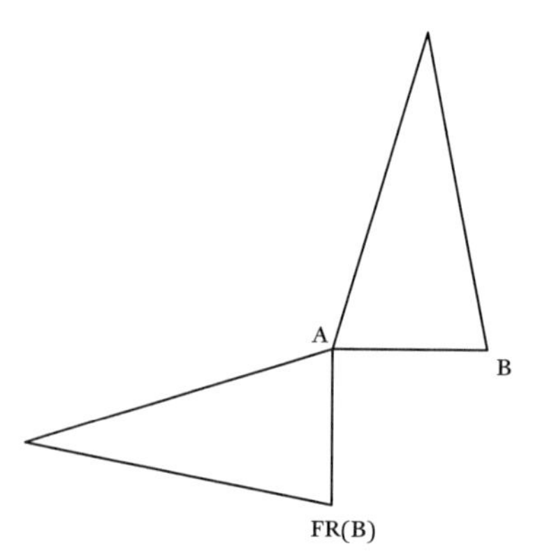

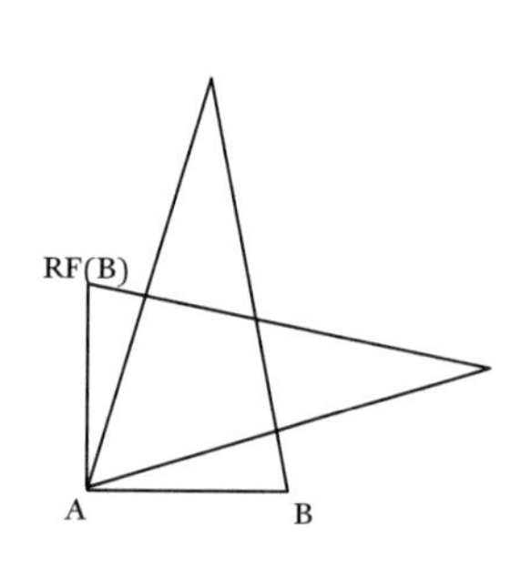

Problem

Suppose **R** *is a rotation about the point P through an angle of* 30° *and* **R'** *is a rotation through an angle of* 40° *around a* different *point P'. What type of isometry is the isometry* **R'R**? *Why? (HINT: Draw a geometric object and see what happens to it.)*

Problem

Suppose **F** *is a flip across the line* ℓ *and* **F'** *is a flip through a* different *line* ℓ'. *What type of isometry is the isometry* **F'F**? *Why? (HINT: Draw the two lines* ℓ *and* ℓ' *and look at where they intersect (if they do intersect!).)*

Problem

Make up your own question similar to the last two. Figure out the answer.

The process of putting two isometries **U** and **V** together is called *composition*. The isometries of the plane form an algebraic object called a *group* (see the definition below). The group operation is composition; the fact that **UV** and **VU** are not necessarily equal means that the group is not *commutative*. We will have more on noncommutative groups when we look at the rotations of three-dimensional space in Chapter 5. Because each object **U** in our group is a *transformation*, that is, a rule that sets up a correspondence between points A and $\mathbf{U}(A)$, the group is called a *transformation group*. In the next section we will be interested in collections of isometries that by themselves form transformation groups. The key property is that the composition of any two isometries in our collection should be in our collection.

Here is one simple but important example. If **U** is any isometry, we can form the composition **UU**; call it $\mathbf{U}^2$. We can repeat this and form the composition $\mathbf{U}^2\mathbf{U} = \mathbf{U}^3$. Keep going; we get all the powers $\mathbf{U}^n$, where n is any positive integer. Define the *inverse* of **U**, called $\mathbf{U}^{-1}$, by the rule $\mathbf{U}^{-1}(A') = A$ if $\mathbf{U}(A) = A'$. For instance, if **T** moves everything up one inch, then $\mathbf{T}^{-1}$ moves everything down one inch. If **R** rotates everything 30° around O, then $\mathbf{R}^{-1}$ rotates everything −30° around O. (In other words, it rotates in the opposite direction.) If **F** flips everything across ℓ, then $\mathbf{F}^{-1}$ flips everything back across ℓ. (Hmm. That means that $\mathbf{F}^{-1}$ is the same as **F**!)

Problem
When is $\mathbf{U}^{-1}$ *equal to* **U**?

Let's call $\mathbf{U}^{-1}\mathbf{U}^{-1}$ by the name $\mathbf{U}^{-2}$. Now we can define $\mathbf{U}^n$ for any integer n, positive or negative. What is $\mathbf{U}^0$ supposed to be? In a moment we will see that the answer is that it should be the *identity* **I**, which leaves everything where it is. If we make that definition, then the reward is that the following formula is always true:

$$\mathbf{U}^n\mathbf{U}^m = \mathbf{U}^{n+m}.$$

Notice that this says that if we take two powers of **U** and compose them, we get another power of **U**. This means that the powers of **U** form a group; we will call this group $< \mathbf{U} >$ and call it the *cyclic group generated by* **U**.

Groups. A set of objects G and an operation $\circ$ on pairs of objects forms a group if:

1. Whenever A and B are two objects in G, then $A \circ B$ is an object in G ("CLOSURE").
2. $A \circ (B \circ C) = (A \circ B) \circ C$ ("ASSOCIATIVE LAW").

3. There is a special object I that satisfies the formula $I \circ A = A = A \circ I$ for every A ("IDENTITY").
4. For each object A in G, G contains an object A^{-1} that satisfies the formula $A \circ A^{-1} = I = A^{-1} \circ A$ ("INVERSE").

Problem (The dihedral group D_n)
Suppose $\mathbf{R}$ *is a rotation around a point* O *and* A *is some other point. The points* $A, \mathbf{R}(A), \mathbf{R}^2(A), \mathbf{R}^3(A), \ldots$ *all lie on a circle centered at* O. *If the angle* θ *divides* $360°$ *evenly, so that* $360 = n\theta$, *then these points are the corners of a regular* n*-gon* P. *If we connect* A *to* $\mathbf{R}(A)$ *with a line segment, then* $\mathbf{R}$ *takes that segment to the one joining* $\mathbf{R}(A)$ *to* $\mathbf{R}^2(A)$, *and so on. The rotation* $\mathbf{R}$ *takes the polygon* P *to itself, and so does every* $\mathbf{R}^k$. *Now, although we have infinitely many* names $\mathbf{R}^k$, *there are actually only* n *different rotations:* $\mathbf{R}^n = \mathbf{I}$, $\mathbf{R}^{n+1} = \mathbf{R}$, $\mathbf{R}^{n+2} = \mathbf{R}^2$, *etc. The group* $< \mathbf{R} >$ *is the group of* rotations of the n-gon, *and we will call it* C_n. *But these are not the only isometries that take* P *to itself. The flip* $\mathbf{F}$ *across the line through* O *and* A *also takes* P *to itself. Take any other vertex* V *of* P *and there will be a flip across the line through* O *and* V *that takes* P *to itself. If we add all of these flips to the rotations, we get a group of isometries containing exactly* $2n$ *isometries. This is the dihedral group* D_n.

How can we see that there are exactly $2n$ isometries that take P to itself? Any such isometry must take A to one of the n vertices of P. So we need only see that there are exactly two isometries that take A to a given vertex V. But the vertex $\mathbf{R}(A)$ must go to one of the two vertices adjacent to V, and after that there is no more choice.

Since $\mathbf{RF}$ takes P to itself, it must be one of the isometries described above. In fact, it is one of the flips. (Which one?) Every transformation in D_n can be described either as being in C_n or as $\mathbf{R}^k\mathbf{F}$ for some k.

Problem
Find the rule for changing $\mathbf{FR}^j$ *into something of the form* $\mathbf{R}^k\mathbf{F}$. *Does* k *ever equal* j?

2.2 Regular and Semiregular Tessellations

A *regular tessellation,* or *regular tiling,* of the plane is made by taking identical copies of a regular polygon P_n with n sides and covering every point in the plane so that there is no overlap except for the edges: Each edge of a polygon coincides with the edge of one other polygon. It is pretty easy to discover that there are just three types of regular tessellation. This is a somewhat disappointing state of affairs, which turns out to be the fault of the parallel postulate. We will fix this in the next chapter.

In order to make a regular tessellation, we have to be able to fit a certain number of copies of a polygon together so that they share a common vertex. This means that the interior angle of the polygon has to divide 360° evenly. Let's compute the interior angles of an n-gon. If we draw all the diagonals from one vertex to the other vertices, we cut the polygon up into exactly $n-2$ triangles. The sum of the angles in each one is 180°. So the sum of the interior angles of all the triangles is $180(n-2)°$. But this is the same as the sum of the interior angles of the polygon (draw a picture to see why). Since all the angles are the same, the angle at each vertex is $\frac{180(n-2)}{n}$ degrees. This will divide 360 evenly only if $360 \times \frac{n}{180(n-2)} = \frac{2n}{n-2}$ is a whole number. So n must be 3, 4, or 6. (After 6 the fraction becomes smaller than 3 but must stay larger than 2.) Each of the three possibilities gives rise to one of the regular tessellations.

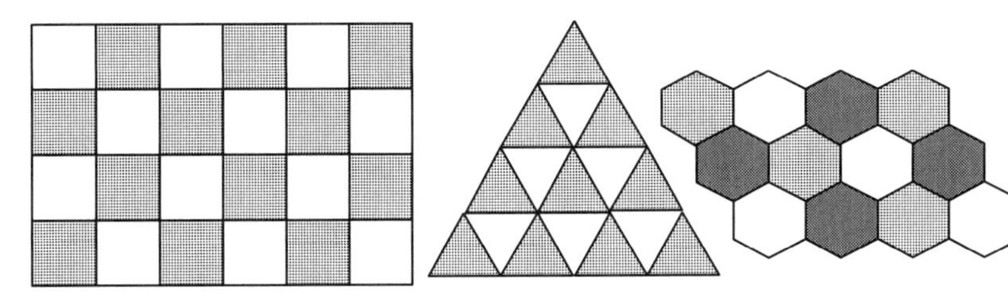

The three regular tessellations

The three familiar patterns each possess a large amount of *symmetry*. What this means is that for each pattern there is a group of isometries that carries each polygon in the pattern onto a polygon in the pattern. For instance, the tiling of the plane by squares of unit side length is carried to itself by the translation **H** that slides everything to the right one unit, and also by the translation **V** that slides everything up one unit. Pick one square and call it the "home" square. Let the dihedral group D_4 (see the last section) be the group of isometries that take the home square to itself; then these isometries also preserve the pattern. For example, if **R** is the rotation by 90° around the center of the home square, then the whole pattern rotates into itself. Any composition of symmetries of the pattern will again be one; for instance, we can take $\mathbf{V}^{-3}\mathbf{H}^5$, which moves every square five units to the right and three down.

Now in fact, that is the complete list of symmetries for this pattern. To see this, we first notice that we can move the home square to any other square by some combination of horizontal and vertical moves. If we *first* rotate or flip the home square and then move it to another square, we can find eight different ways to move the home square to each other square. But those are the only possible ways to move the home square to another square, and once we know what has happened to the home square, we know where everything else must move.

The translations that preserve the pattern are *generated* by **H** and **V**. (That means that every translation can be made by repeated use of **H**, **V**, and their inverses.) Since the composition of translations is always a translation, there is a general principle at work here: If we have any group G of isometries of the plane, then the translations that are in G form a group K by themselves. Since every isometry in K is in G, we say that K is a *subgroup* of G. We write $<\mathbf{H}, \mathbf{V}>$ for the group of translations of the square tiling. If P is any point in the plane, then the set Γ of points $\mathbf{T}(P)$, where **T** ranges over all the translations in the group, is a (square) *lattice*.

We can go through the same procedure to find the group of symmetries of each of the other patterns. In the case of the triangular tiling, however, the story is a bit more complicated. We have colored the pattern with black and white triangles. Let's suppose that the home triangle Q is white. Then a translation can take the home triangle only to one of the white triangles. The translation **H** that moves one unit to the right and the translation **D** that moves diagonally (at a 60° angle) up and to the right one unit generate the group $<\mathbf{H}, \mathbf{D}>$ of translational symmetries. The lattice Γ of translates of a starting point P is made up of the corners of the triangles. It is called a *rhombic lattice,* since there is a rhombus (made up of two adjacent triangles) whose translates by members of this group fill out the plane.

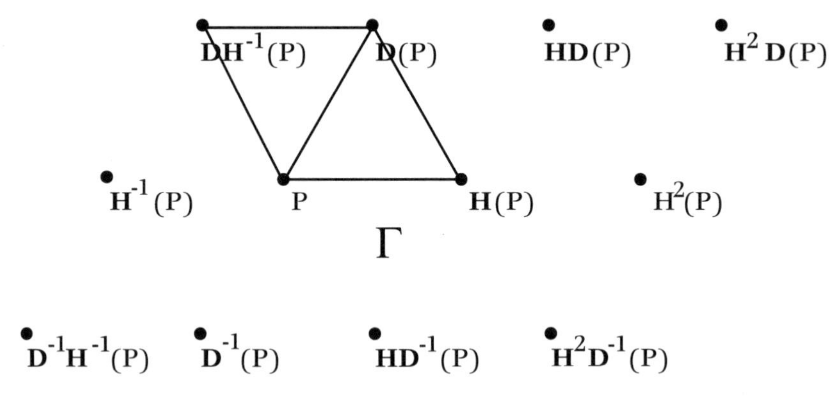

The dihedral group D_3 acting on the home triangle also carries triangles to those of the same color. Let **R** be the 180° rotation about the lower vertex P of the home triangle. (**R** is called a *half-turn*). Unlike **H** and **D**, **R** takes white triangles to black triangles. Using **H**, **D**, and **R**, it is possible to take Q to any other triangle in the plane. Consequently, we can conclude that **R** together with **H**, **D**, and D_3 *generate* the symmetries of the triangular lattice. That means that every symmetry is made up of a composition of these isometries.

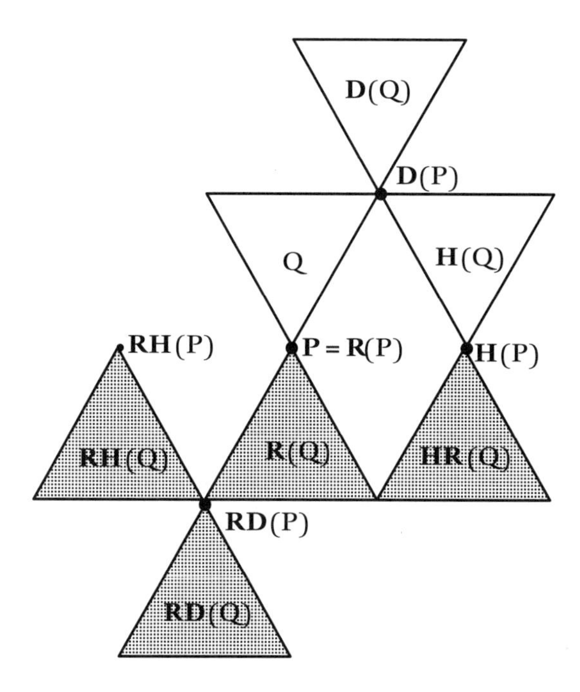

Here is another description of the group. Let **S** denote the rotation through 60° around the lower vertex P of the home triangle. Let **F** be the flip across the line through the base of the home triangle. Then the symmetries of the lattice are generated by **H**, **D**, **F**, and **S**. To see this, we can check that $\mathbf{HS}^2$ rotates the home triangle 120°, while **SF** flips the home triangle keeping P fixed. So we can get every symmetry of the home triangle using **S**, **H**, and **F**. Using **D**, **H**, and **F** we can move the home triangle to any other triangle.

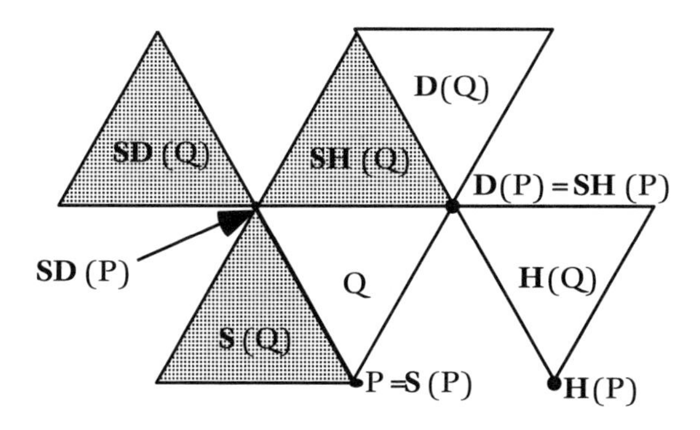

Problem
Show that we don't actually need to use **D**. *So the symmetries of the triangular lattice are generated by* **H**, **F**, *and* **S**.

Problem
Show that we don't actually need to use **H**. *So the symmetries of the triangular lattice are generated by* **D**, **F**, *and* **S**. *[HINT: Look at the isometry* $\mathbf{S}^{-1}\mathbf{DS}$.*]*

We won't go through the analysis for the hexagonal tiling. In fact, we don't have to, because the group of symmetries is just the same group as the one for triangles! There is an easy, and very important, way of seeing this. Take the centers of the hexagons and connect the centers of any two adjacent hexagons by drawing the line segment between them.

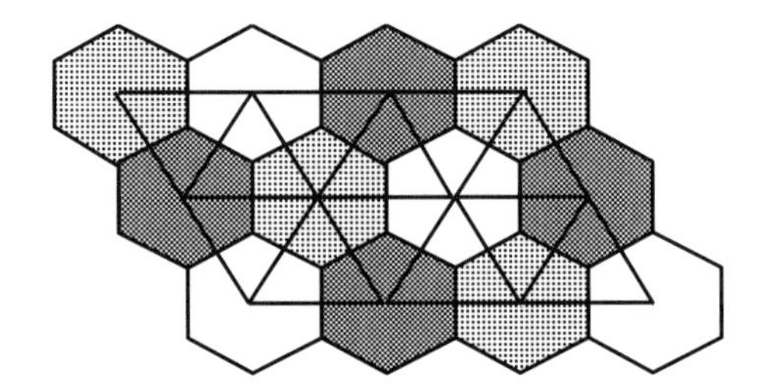

This is the process of constructing the *dual tiling* to the tiling we started with. Since all the hexagons are the same, the centers of adjacent ones are all at the same distance from each other. So the dual tiling is just the regular triangular tiling. Since any symmetry of the hexagonal tiling must take the center of each hexagon to the center of a hexagon, the symmetries also are symmetries of the dual tiling, and vice versa.

If we do the same construction for the triangular tiling, we get the hexagonal tiling. If we do the dual construction for the square tiling, we get another copy of the square tiling.

If we want a tiling of the plane by regular polygons to have the property that there is a symmetry carrying any polygon to any other polygon, then all the polygons have to be the same. They don't have to be regular polygons, though. For instance, we have already seen an example (slightly hidden) of a tiling of the plane by copies of a rhombus. More on this later.

Meanwhile, suppose we insist on regular polygons but allow more than one kind of polygon. Then there is no chance for symmetries carrying a fixed polygon to every other one, so instead let's ask that there are symmetries carrying a particular *vertex* P to every other vertex. A moment's thought reveals that in that case there have to be the same number of polygons meeting at each vertex. Also, if we start at one vertex and walk around it in a circle, we should encounter a certain sequence of polygons, with the same sequence occurring when we walk around any other vertex. For instance, we might see a square, then an octagon, then another octagon, and then be back where we started. This is actually possible, because an octagon has 135° angles. (Remember, the formula is $\frac{180(n-2)}{n}$.) For the purpose of keeping track of patterns, let's give this pattern the name $(4, 8, 8)$. In general, the name of a pattern tells us the number of sides in each polygon we encounter as we walk (counterclockwise) once around each vertex. Of course, the same pattern has more than one name; for instance, we could call this pattern $(8, 4, 8)$.

What patterns are possible? To search systematically, we consider patterns with three polygons at each vertex, then four, then five, and then six. Then we will be done, because the smallest angle possible is 60° (in a triangle), so there can't be more than six angles coming together at a corner.

Three at a vertex. If there are three polygons at each vertex, then we have a pattern (n_1, n_2, n_3) where

$$180\frac{n_1 - 2}{n_1} + 180\frac{n_2 - 2}{n_2} + 180\frac{n_3 - 2}{n_3} = 360,$$

or

$$\frac{n_1 - 2}{n_1} + \frac{n_2 - 2}{n_2} + \frac{n_3 - 2}{n_3} = 2.$$

A little algebra turns this into the equation

$$\frac{1}{n_1} + \frac{1}{n_2} + \frac{1}{n_3} = \frac{1}{2} \quad . \tag{2.2.1}$$

There are ten sets of integers that work. The first, $(6, 6, 6)$, comes from the regular hexagonal tiling. Three others are $(12, 12, 3)$, $(12, 4, 6)$, and $(8, 8, 4)$. Each of these turns out to generate a pattern that fills the whole plane. These tilings are called *semiregular tilings*.

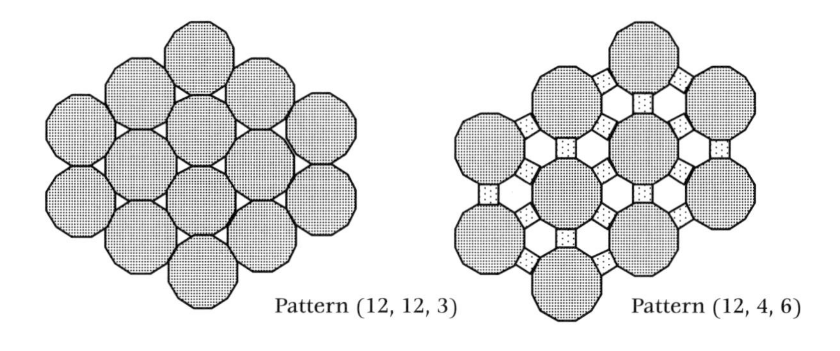

Pattern (12, 12, 3)   Pattern (12, 4, 6)

Problem
Find the other six patterns. Figure out why none of them can be used to generate semiregular tilings.

Four at a vertex If there are four polygons at each vertex, then we have a pattern (n_1, n_2, n_3, n_4). A similar calculation leads to the formula

$$\frac{1}{n_1} + \frac{1}{n_2} + \frac{1}{n_3} + \frac{1}{n_4} = 1 \quad . \tag{2.2.2}$$

Problem
Find all allowable patterns with four polygons at each vertex. For each one, try to build a semiregular tiling of the plane.

Five at a vertex If there are five polygons at each vertex, then we have a pattern $(n_1, n_2, n_3, n_4, n_5)$. This time the formula is

$$\frac{1}{n_1} + \frac{1}{n_2} + \frac{1}{n_3} + \frac{1}{n_4} + \frac{1}{n_5} = \frac{3}{2} \quad . \tag{2.2.3}$$

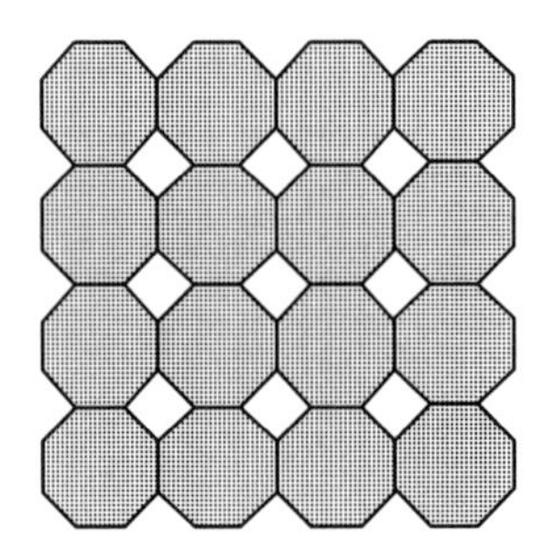
Pattern (8, 8, 4)

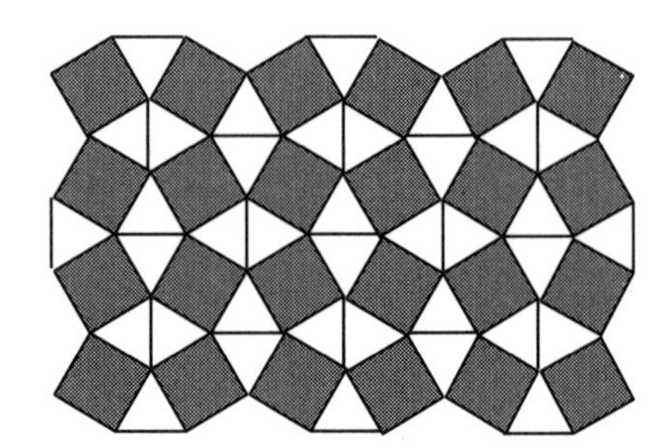
Pattern (4, 3, 4, 3, 3)

How do we solve (2.2.3)? Of the five numbers, the smallest one cannot be bigger than 3, since the total is $\frac{3}{2}$. So let $n_5 = 3$. The remaining fractions therefore sum to $\frac{7}{6}$, so one of them must again be $\frac{1}{3}$. The remaining three fractions sum to $\frac{5}{6}$, so again one of them must be $\frac{1}{3}$. Now we need to solve $\frac{1}{a} + \frac{1}{b} = \frac{1}{2}$. This has two solutions: $a = 3,\ b = 6$ and $a = 4,\ b = 4$. This gives the following different patterns: $(3, 3, 3, 4, 4)$, $(4, 3, 4, 3, 3)$, and $(6, 3, 3, 3, 3)$. This gives three more semiregular tilings.

The patterns with four polygons do not all extend to the whole plane, as you may have already discovered. The two that do are pictured below. They bring the total number of semiregular tilings up to eight. Interestingly, one of the patterns is different from its mirror image. If you make copies of all eight patterns on transparencies, and then flip the transparencies over, all the others will fit over themselves. In other words,

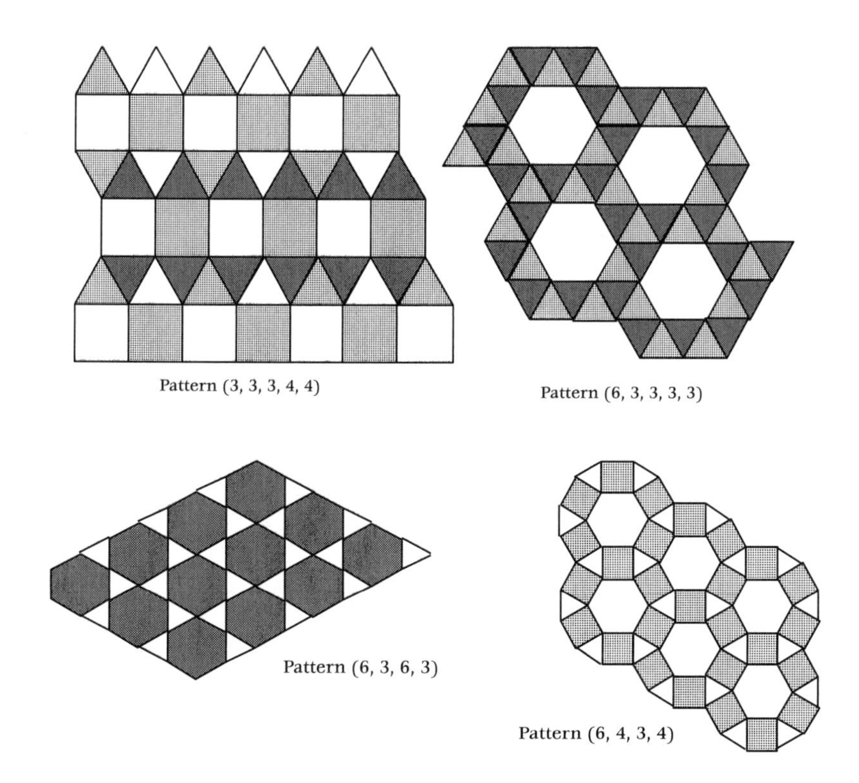

Pattern (3, 3, 3, 4, 4)

Pattern (6, 3, 3, 3, 3)

Pattern (6, 3, 6, 3)

Pattern (6, 4, 3, 4)

the group of symmetries of one of these eight patterns does not contain a flip or glide reflection. (Can you see which one it is?)

Problem

Look at the symmetries of each of the semiregular tilings. Check that there is a symmetry taking each vertex to any other vertex. Obviously a symmetry must take a polygon to one with the same number of sides. Which patterns (if any) have congruent polygons that can not *be taken to each other by a symmetry of the tiling?*

We constructed the duals of the regular tilings; these turned out to be regular tilings. The duals of semiregular tilings do not turn out to be semiregular tilings. Since we connect the centers of adjacent polygons, we get a polygon around each vertex of the original tiling. In the case of semiregular tilings, these polygons are all congruent to each other. But they are not regular polygons. Actually, that's good news, because it gives us eight pretty examples (some prettier than others) of tilings by congruent polygons. In particular, it is possible to fill up the plane with congruent pentagons. The symmetries of this tiling carry any pentagon to any other pentagon.

Problem
Determine the edge lengths and angles of a pentagon that tiles the plane in this way.

One particularly elegant tiling of the plane by pentagons is known as the *Cairo tessellation*, because it can be seen as a street tiling in Cairo. The pentagon used for this tiling can be constructed using straightedge and compass; in fact, the pentagon in the picture below was constructed using *the Geometer's Sketchpad*,[1] a program with wonderful construction tools. Although it is not regular, it is *equilateral*, that is, all the sides have the same length.

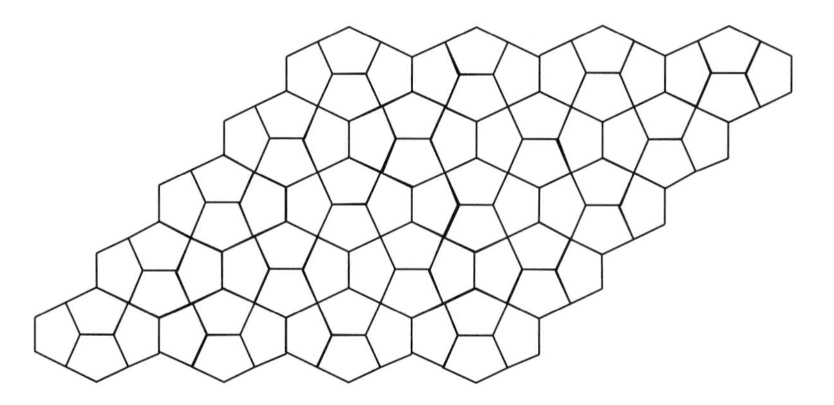

Tiling the plane with identical copies of any rhombus is easy: Just modify the usual pattern for squares to make a slanted checkerboard. In fact, this works for any parallelogram. That means that if we start with any triangle, then we can tile the plane with copies of it by putting pairs of them together to make parallelograms. The symmetries of each tiling carry any polygon to any other one. Among the dual tilings to the semiregular tilings, there is one in which the polygons are rhombi. Unlike the checkerboard tilings, this pattern does not have the same number of rhombi coming together at each vertex; but it is still possible to find an isometry carrying any rhombus to any other.

The problem of tiling the plane with pentagons has a fascinating history. A good place to read about it is in Martin Gardner's book [17], chapter 13. You might also read [14], chapters 1 and 2, which deal with tilings that are *not* symmetric. A thorough discussion of symmetry and tilings is the book [26] by George Martin. Lastly, I would like to mention the picture book [5] as a source of artistic inspiration.

---

1. Key Curriculum Press, Berkeley, 1994.

2.3 Tessellations That Aren't, and Some Fractals

In constructing regular and semiregular tilings, the primary requirement was that the resulting pattern have a large symmetry group. In a semiregular tessellation, there is an isometry of the plane carrying any vertex to any other vertex. A tiling with this property is called *vertex transitive.* A tiling is called *tile transitive* if there is an isometry carrying any tile to any other tile. Among the semiregular tilings, only the regular ones are tile transitive.

Problem

A tiling is called edge transitive *if there is an isometry carrying any edge to any other. Which semiregular tilings have this property?*

In the last section, we ran into some patterns that did not produce semiregular tessellations. My personal favorite is the pattern (10,5,5). The interior angles in a regular pentagon are 108°, while those of the decagon are 144°. That implies that if we take ten regular pentagons and put them in a ring, they will fit together perfectly around a decagon. Unfortunately, when we try to enlarge this pattern, we are forced to put three pentagons together at one corner, where they leave a little gap (36°). This prevents us from tiling the plane with regular decagons and hexagons, although it does not prevent the creation of beautiful designs.

The greatest monument to geometrical design is the Alhambra, in Granada, Spain. Built in the thirteenth century, when Granada was the capital of Moorish Spain, it is decorated with intricate designs of a wide variety of geometric patterns. The Islamic religion forbids the graphic representation of living things, and this prohibition was rigorously applied. (Judaism also has such a prohibition: See Exodus 20:4, the second

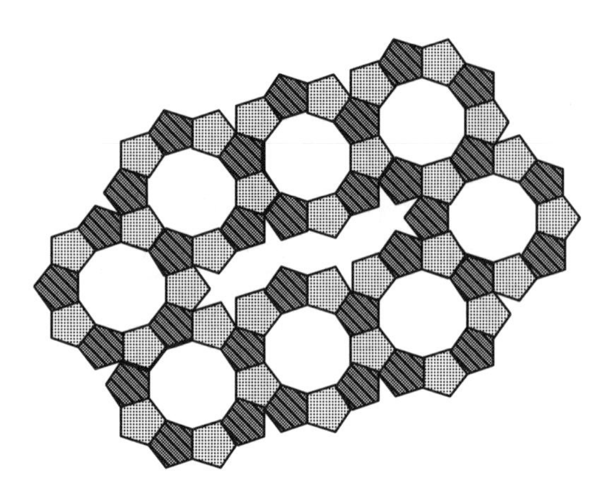

Commandment. The Second Council of Nicea in 787 ruled that this did not apply to Christians, overruling the "Iconoclasts," who wanted to apply this commandment rigidly.) Islamic artists dealt with this prohibition when decorating mosques by concentrating on nonrepresentational design. Clearly, the symmetry of a pattern was to them a reflection of its spirituality.

A consequence of this is that the numbers 3, 4, 6, 8, 12, and 16 appear in great abundance in the designs found in the Alhambra, while 5 and 10 are not very common. There are examples of symmetrical patterns in Moorish art that contain pentagons, decagons, and five- and ten-pointed stars. See plates 171–190 in [5]; one example is sketched below. There are even a few designs containing 7 and 14 (see plates 164-170). Still, one can search for hours through a bewildering array of intricate designs in the Alhambra without encountering many such objects. (I was unable to locate an example of a seven-sided figure during a recent visit.)

Problem

Even this pattern of pentagons and ten-pointed stars does not exhibit fivefold symmetry. Find the symmetries of this pattern. If S is a symmetry of the pattern for which $S^n = I$ for some integer n, what values can n take?

We shall now examine a geometric object with a new kind of symmetry. A *similarity* of the plane is a transformation such that for some fixed constant r, it carries every line segment AB to a line segment $A'B'$ whose length $|AB| = r|A'B'|$. If $r = 1$, then this is just an isometry. When $r \neq 1$, a triangle ABC is taken to a triangle $A'B'C'$ that is similar but not congruent. (Recall from Chapter 1 that Wallis's alternative to the parallel postulate says that there exist similar, noncongruent triangles. So this is definitely a Euclidean concept.) Such a transformation is called a *dilation*.

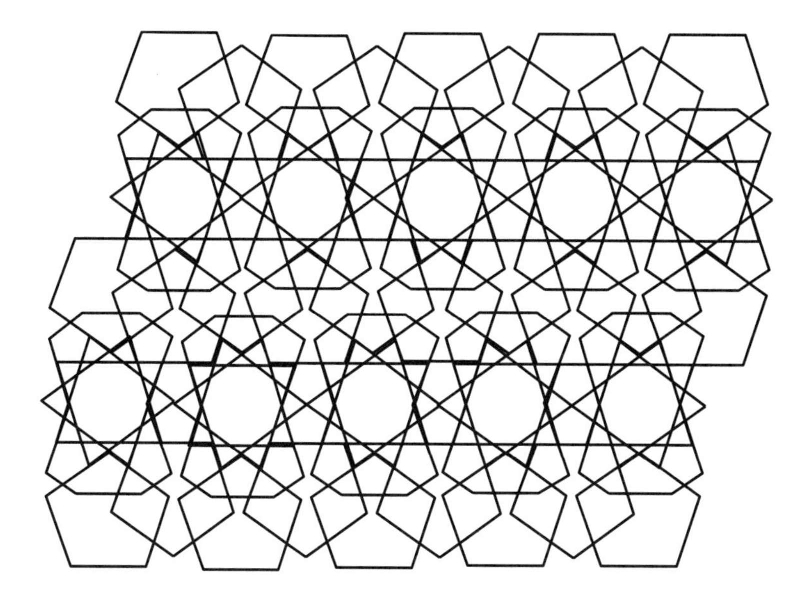

A regular pentagon can be divided up into six regular pentagons and four small triangles. Each of the small pentagons is similar to the original large pentagon. Therefore, we may divide each of those six pentagons into six smaller similar pentagons (and four left-over triangles). This gives a pleasing array of 36 tiny pentagons. You may also observe a few decagons almost but not quite appearing in the design.

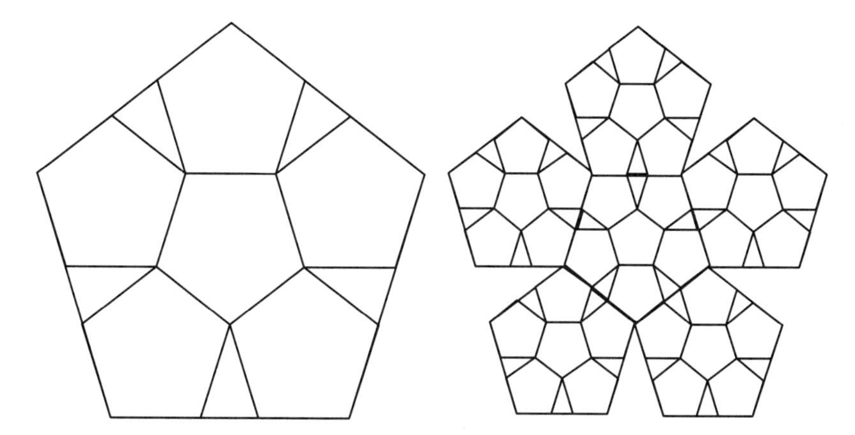

Now, if we take one of the medium-sized pentagons in this picture, it contains a design within it that looks *somewhat* similar to the overall design. The difference is that it lacks the fine detail: the larger pentagonal "snowflake" has 36 little pentagons in it, while the smaller one has only six. To fix that, replace each tiny pentagon by a snowflake of six smaller pentagons. This leads to a more elaborate pattern, consisting of a large "snowflake" made up of six smaller "snowflakes" each of which is almost similar to the large one.

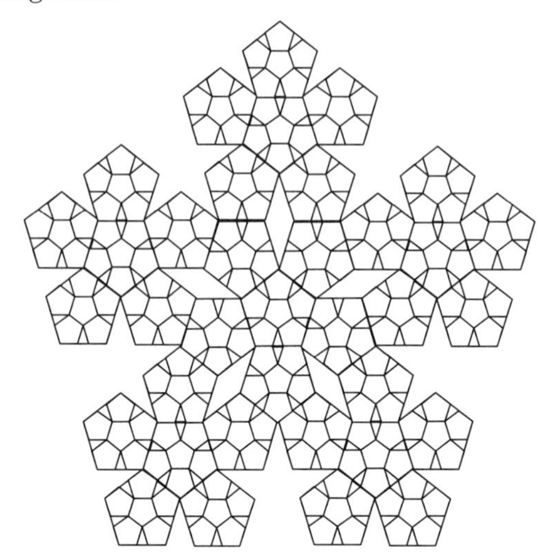

Imagine that we could continue this process indefinitely, leading to more and more complicated patterns. The result of this process would be a geometric object that has the same isometry group as the pentagon (the dihedral group D_5) but also has an additional type of symmetry. We can find arbitrarily tiny pieces of it that are *exactly* similar to the whole. Such an object is called a *fractal*. I like to call this particular example a "pentaflake." *Snowflake* is not really an appropriate term, since real snowflakes have six-fold symmetry.

It is also possible to fill out the entire plane by this process (except that the funny-shaped holes will get larger and larger). Starting with the pentaflake, construct an array of six of them; this fills out a larger pentagonal region of the plane. Repeat this process indefinitely. In the end, we will have a pattern with the weird property that there is a dilation that carries the pattern exactly onto itself!

Fractals. The term *fractal* was coined by the mathematician Benoit Mandelbrot [25] to describe a geometrical object with the property that no matter how closely you look at it, it always looks the same. Such an object is called *self-similar*. (Actually, that definition can be loosened to "approximately self-similar," meaning that small pieces look roughly similar to the whole. A famous example of such an object is the *Mandelbrot set*, which arises in the study of iteration of functions.)

The oldest example of such a geometric object is the *Koch snowflake*, constructed by the Swedish mathematician Helge von Koch in 1904. The procedure for constructing the snowflake is simple. We begin with an

equilateral triangle. The first stage places a smaller equilateral triangle on the middle third of each side of the triangle to produce a six-pointed star.

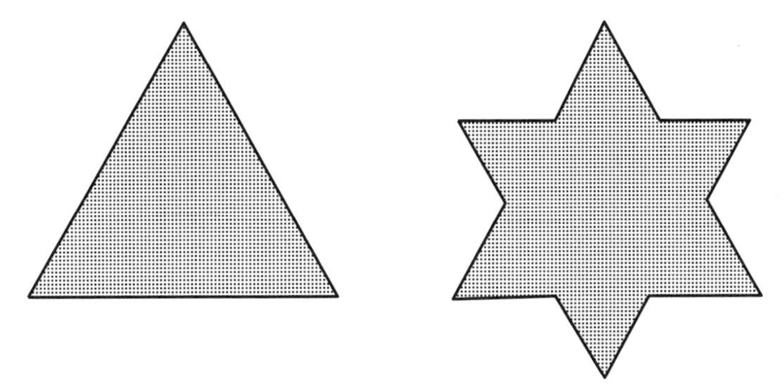

Each successive stage is produced by putting an equilateral triangle on the middle third of each side of the preceding figure. In the end we get a snowflake. To be precise, the Koch snowflake is actually the boundary curve. It has some rather remarkable properties. For example, suppose we start with a triangle of side 1. Then its perimeter is 3. The six-pointed star has perimeter 4, because each side of the triangle has been replaced by a curve that is $\frac{4}{3}$ as long. Now, at the next stage each edge of the star is replaced by a curve that is $\frac{4}{3}$ as long, so the new figure has perimeter

$\frac{4}{3} \times \frac{4}{3} = \frac{16}{9}$. The next figure (pictured to the right below) has perimeter $\frac{64}{27}$. Continuing in this way, we get figures whose perimeters are $\left(\frac{4}{3}\right)^n$. When n gets big, these numbers also get big. (Since $\left(\frac{4}{3}\right)^4 > 3$, we can see that $\left(\frac{4}{3}\right)^{4n} > 3^n$.) So the Koch snowflake has *infinite perimeter!*

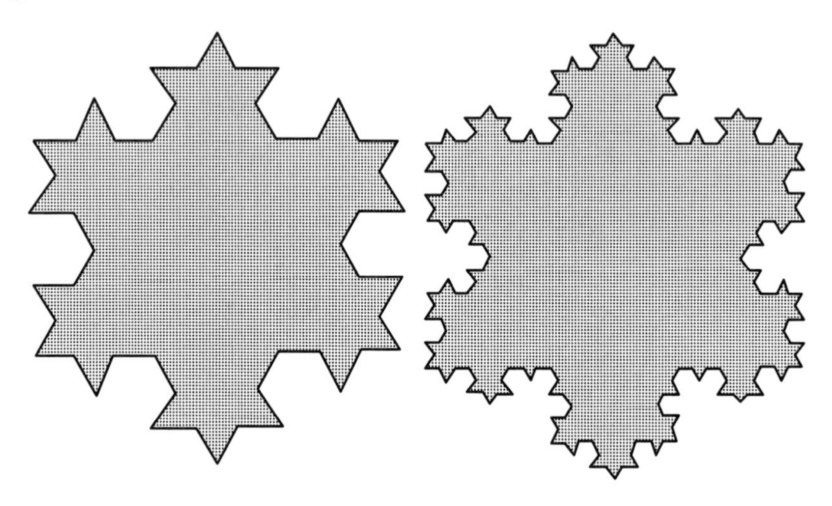

On the other hand, it is pretty clear that the areas of the successive stages do not grow very much. It can be shown that the area enclosed by the Koch snowflake curve is exactly $\frac{2\sqrt{3}}{5}$.

This strange behavior is characteristic of fractal curves. One way to see that the perimeter of the curve is infinite is to use the property of self-similarity. Suppose the perimeter were exactly $3x$, where x is the length of one of the three "sides" of the snowflake. (That is, what evolved from one of the sides of the original triangle.) Well, that side is made up of four identical pieces, each of which must therefore have length $\frac{x}{4}$. But if we take one of those pieces and dilate it by a factor of 3, it will exactly coincide with the whole side. That means that $3 \times \frac{x}{4} = x$. Uh-oh! This can't be true, so the assumption that the number x even exists must be false.

There is another way of interpreting this last argument. If we triple the dimensions of a rectangle, its area increases by a factor of nine, while its perimeter only increases by a factor of three. The same is true for a triangle or a circle. If we triple the dimensions of a cube, the volume increases by a factor of $27 = 3^3$, and the same is true for any other solid object. This is one way of distinguishing between "two-dimensional" and "three-dimensional" objects. If we take a side of the Koch snowflake curve and triple its "dimensions," it becomes $4 = 3^{1.262}$ times as large. By this reasoning, the snowflake curve has dimension 1.262 ($= \log 4/\log 3$). It has *fractional dimension*, whence the name fractal. Mandelbrot observed that various objects in nature such as geographical coastlines and mountain ranges exhibit this fractional behavior.

One further example of a fractal is constructed by subdividing an equilateral triangle into four smaller ones and removing the middle one (which is upside down). This process is then repeated on the remaining smaller triangles. A similar analysis to that of the Koch snowflake shows that in the end, the area of the resulting geometric object is 0. This is an indication that this object is not really two-dimensional, but that it has a fractional dimension less than two.

Problem

What is the dimension of this self-similar triangular figure? What is the dimension of the pentaflake?

A pleasant introduction to fractals can be found in chapter 3 of [14]. A mathematical treatment of fractals and fractional dimensions is contained in the book [12], and the applications of fractal geometry are explored in Mandelbrot's book [25], which features some spectacular pictures.

The Renaissance artist Albrecht Dürer, who was fascinated by tilings, produced a tiling of the plane using regular pentagons with diamond-shaped gaps. Here is a version of his tiling; note that it possesses only fivefold symmetry.

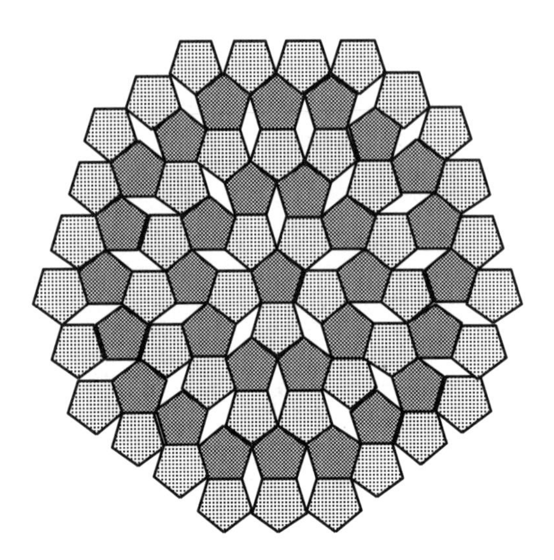

2.4 Complex Numbers and the Euclidean Plane

It is a theorem of Euclidean plane geometry that every isometry is either a translation, a rotation, a reflection, or a glide reflection. This is not entirely obvious; in fact, in non-Euclidean geometry it is false! So to the extent that our intuition about geometry does not include understanding parallel lines, we should not expect to have an intuitive understanding of the nature of isometries. In particular, if we compose two rotations about different points it is not so easy to see what the result is.

In this section we will begin using a powerful computational tool to analyze isometries: complex arithmetic. In Chapter 3 we will use this tool to study isometries of the hyperbolic plane. In Chapter 5 we will use a generalization of complex numbers to study rotations of the sphere.

Analytic geometry translates statements about points and lines into statements about ordered pairs of numbers and equations. We will carry this one step further by thinking about ordered pairs as numbers that we can add, subtract, multiply, and divide.

We begin by assuming that there is a "number" i (i for imaginary) that when multiplied by itself gives -1. Supposing that there is such a number, with which we can do arithmetic along with ordinary numbers, we can then form numbers of the form $\alpha = a + bi$, called *complex numbers*. We will do arithmetic with these numbers by using the ordinary rules of arithmetic and remembering that when i is multiplied by itself it yields -1. Geometrically, $a+bi$ is just the ordered pair (a,b).

To add $\alpha = (a+bi)$ to $\beta = (c+di)$, we must use several rules of arithmetic. Let's do this carefully:

$$\begin{aligned}(a+bi)+(c+di) &= ((a+bi)+c)+di && \text{(Associative Law)}\\ &= (a+(bi+c))+di && \text{(Associative Law)}\\ &= (a+(c+bi))+di && \text{(Commutative Law of addition)}\\ &= ((a+c)+bi)+di && \text{(Associative Law)}\\ &= (a+c)+(bi+di) && \text{(Associative Law)}\\ &= (a+c)+(b+d)i && \text{(Distributive Law).}\end{aligned}$$

This process did not use anything about the number i. In fact, if we wanted to add ordered pairs of numbers, we could just write

$$(a,b)+(c,d) = (a+c, b+d).$$

This is the usual definition of *vector addition*. Geometrically, adding complex numbers is just like adding vectors.

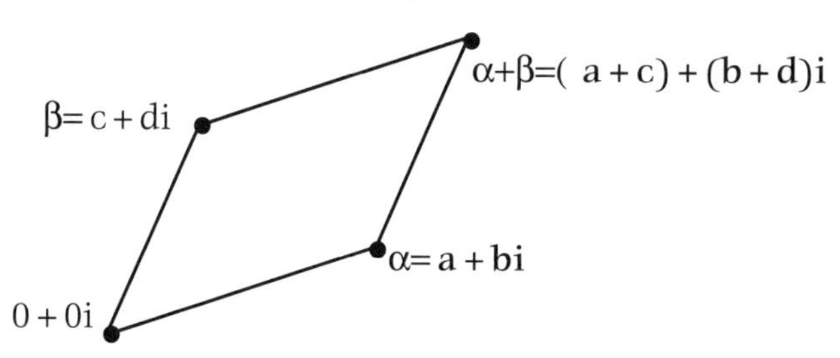

Subtraction works the same way as subtraction of vectors. Multiplication is more interesting.

$$\begin{aligned}\alpha\times\beta &= [(a+bi)\times c]+[(a+bi)\times(di)] && \text{(Distributive Law)}\\ &= [ac+((bi)\times c)]+[(a+bi)\times(di)] && \text{(Distributive Law)}\\ &= [ac+((bi)\times c)]+[(a\times(di))+((bi)\times(di))] && \text{(Distributive Law)}\\ &= [ac+((bi)\times c)]+[((ad)i)+((bi)\times(di))] && \text{(Distributive Law)}\\ &= [ac+(b(i\times c))]+[((ad)i)+((bi)\times(di))] && \text{(Associative Law)}\\ &= [ac+(b(i\times c))]+[((ad)i)+(b(i\times(di)))] && \text{(Associative Law)}\\ &= [ac+b(ci)]+[((ad)i)+(b(i\times(di)))] && \text{(Commutative Law)}\\ &= [ac+(bc)i]+[((ad)i)+(b(i\times(di)))] && \text{(Associative Law)}\\ &= [ac+(bc)i]+[((ad)i)+(b((id)i))] && \text{(Associative Law)}\\ &= [ac+(bc)i]+[((ad)i)+(b((di)i))] && \text{(Commutative Law)}\end{aligned}$$

$$
\begin{aligned}
&= [ac + (bc)i] + [((ad)i) + (b(d(i \times i)))] && \text{(Associative Law)} \\
&= [ac + (bc)i] + [((ad)i) + (b(d(-1)))] \\
&= [ac + (bc)i] + [(ad)i - bd] \\
&= (ac - bd) + (bc + ad)i.
\end{aligned}
$$

It is important to note that we needed to use the fact that $i \times d = di$. This is the commutative law of multiplication, but since i is not an ordinary number, we needed to adopt that rule in order to complete the calculations. This little point will become significant in Chapter 5.

Division of complex numbers poses more of a challenge than other operations. To divide α by β, we need to find a complex number γ that when multiplied by β gives us back α. If $\beta = c$ is a real number, this is easy. We have $(\frac{a}{c} + \frac{b}{c}i) \times c = a + bi$. This makes sense, provided that $c \neq 0$. Now, by a very clever trick we may reduce the problem of division to this special situation. First we need a definition.

If $\alpha = a + bi$ is a complex number, then its *complex conjugate* $\bar{\alpha}$ is the complex number $a - bi$. The real number a is called the *real part* $\mathbf{Re}(\alpha)$ of α, while the real number b is called the *imaginary part* $\mathbf{Im}(\alpha)$ of α. Using complex conjugates, we can find the real and imaginary parts of α by the formulas $2\mathbf{Re}(\alpha) = \alpha + \bar{\alpha}$ and $2\mathbf{Im}(\alpha) = \alpha - \bar{\alpha}$.

Lemma 2.4.1
If α and β are complex numbers, then $\bar{\alpha} + \bar{\beta} = \overline{\alpha + \beta}$, and $\bar{\alpha}\bar{\beta} = \overline{\alpha\beta}$. The product $\alpha\bar{\alpha}$ is real and positive for any $\alpha \neq 0$.

Now we are able to divide complex numbers. If $\alpha = a+bi$ and $\beta = c+di$, then

$$
\frac{\alpha}{\beta} = \frac{1}{\beta\bar{\beta}}\alpha\bar{\beta} = \frac{ac + bd}{c^2 + d^2} + \left(\frac{bc - ad}{c^2 + d^2}\right)i.
$$

If $\alpha = a + bi$, then $\alpha\bar{\alpha} = a^2 + b^2$, which by the Pythagorean theorem is the square of the length of the line segment from $(0, 0)$ to (a, b). We define the *absolute value* of α to be the quantity $|\alpha| = \sqrt{\alpha\bar{\alpha}}$. If α and β are two complex numbers, then $|\alpha - \beta|$ is the length of the line segment joining α to β. If α' and β' are two other numbers in the complex plane, then the line segments joining α to β and α' to β' are congruent if $|\alpha - \beta| = |\alpha' - \beta'|$.

Suppose ρ is a complex number that has absolute value 1. Define a transformation $\mathbf{R}$ of the plane by $\mathbf{R}(\alpha) = \rho\alpha$. Then we have

$$
\begin{aligned}
\left|\mathbf{R}(\alpha) - \mathbf{R}(\beta)\right|^2 &= |\rho\alpha - \rho\beta|^2 \\
&= (\rho\alpha - \rho\beta)(\overline{\rho\alpha - \rho\beta}) \\
&= \rho(\alpha - \beta)(\bar{\rho})(\overline{\alpha - \beta})
\end{aligned}
$$

$$= \rho\bar{\rho}(\alpha - \beta)(\overline{\alpha - \beta})$$
$$= (\alpha - \beta)(\overline{\alpha - \beta})$$
$$= |\alpha - \beta|^2 .$$

In other words, the distance from $\mathbf{R}(\alpha)$ to $\mathbf{R}(\beta)$ is the same as the distance from α to β. This calculation shows that $\mathbf{R}$ is an isometry of the plane. Since for $\rho \neq 1$, $\mathbf{R}(\alpha) = \alpha$ only for $\alpha = 0$, $\mathbf{R}$ must be a rotation around the origin. Using a little trigonometry, we can determine the angle of rotation. Since $|\rho| = 1$, the segment from the origin to ρ has length 1 and makes an angle θ with the positive x-axis. So the coordinates of ρ can be written as $\cos\theta$ and $\sin\theta$. Since $\mathbf{R}(1) = \rho$, $\mathbf{R}$ is rotation by θ. The rotations of the plane around the origin are in 1-to-1 correspondence with the complex numbers ρ satisfying $|\rho| = 1$. If ρ and σ are two such numbers, then their product $\rho\sigma$ satisfies $|\rho\sigma| = |\rho||\sigma| = 1$. The numbers of absolute value 1 make up the *circle group* $\mathbf{S}^1$. The identity element in this group is the number 1. Since $\rho\bar{\rho} = 1$, the complex conjugate of ρ is its inverse in the group. In other words, multiplication by ρ rotates the plane counter-clockwise by an angle θ (assuming that θ is positive), while multiplication by $\bar{\rho}$ rotates the plane clockwise by the same angle.

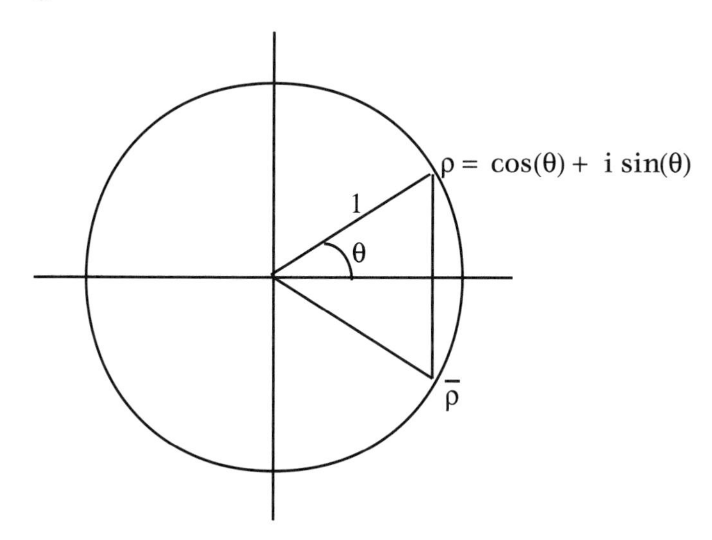

Suppose $\rho = \cos\theta + i\sin\theta$ and $\sigma = \cos\phi + i\sin\phi$ are two members of the circle group. Rotating σ through the angle θ gives us $\cos(\theta + \phi) + i\sin(\theta + \phi)$. Using complex multiplication, $\rho\sigma = (\cos(\theta)\cos(\phi) - \sin(\theta)\sin(\phi)) + i(\sin(\theta)\cos(\phi) + \cos(\theta)\sin(\phi))$. Since these two complex numbers are equal, the real parts must be equal and the imaginary parts must also be equal. These are the *double angle formulas* of trigonometry:

$$\cos(\theta+\phi) = \cos(\theta)\cos(\phi) - \sin(\theta)\sin(\phi),$$
$$\sin(\theta+\phi) = \sin(\theta)\cos(\phi) + \cos(\theta)\sin(\phi).$$

Similarly, we can compute ρ^n two different ways to get *de Moivre's formula*:

$$(\cos\theta + i\sin\theta)^n = \cos(n\theta) + i\sin(n\theta).$$

If α is any complex number other than 0, then multiplication by α is easy to describe. If $|\alpha| = r$, then $\frac{\alpha}{r}$ has absolute value 1. Therefore, we can write $\frac{\alpha}{r} = \cos(\theta) + i\sin(\theta)$, or $\alpha = r(\cos(\theta) + i\sin(\theta))$. Multiplication by a positive real number r just moves every point in the plane to a point r times as far out on a ray from the origin. This is a *dilation* of the plane. (I use this term even when $r < 1$, although its everyday use assumes $r > 1$). Now it is easy to see that multiplication by α is dilation by a factor of r combined with rotation around the origin by angle θ.

If $\eta = a + bi$, then the map $\mathbf{T}(\alpha) = \alpha + \eta$ translates every point in the plane a units horizontally and b units vertically. Subtracting η translates in the opposite direction; it is the inverse $\mathbf{T}^{-1}$ of the translation $\mathbf{T}$. If we want to rotate by an angle θ around the point whose coordinates in the plane are (a, b), then we can translate the plane so that (a, b) moves to the origin, rotate by θ around the origin, and then translate the origin back to (a, b). If $\rho = \cos\theta + i\sin\theta$, and if $\mathbf{R}$ is rotation by θ around the origin, then the rotation $\mathbf{R'}$ around (a, b) is

$$\begin{aligned}\mathbf{R'}(\alpha) &= \mathbf{TRT}^{-1}(\alpha)\\ &= \rho(\alpha - \eta) + \eta\\ &= \rho(\alpha) + \eta(1-\rho).\end{aligned}$$

This is where our work with complex numbers begins to pay off. Any transformation $\mathbf{U}(z) = \rho z + \tau$ of the complex plane with $|\rho| = 1$ is an isometry consisting of rotation by an angle θ followed by translation by the vector (c, d), where $\rho = \cos\theta + i\sin\theta$ and $\tau = c + di$. If $\rho = 1$, then this is a straight translation. If $\rho \neq 1$, then this is a rotation by the angle θ around the point (a, b), where $a + bi = \frac{\tau}{1-\rho}$.

Problem

Suppose $\mathbf{R}$ *is a rotation around* $(1, 0)$ *through an angle of* 30° *and* $\mathbf{R'}$ *is a rotation through an angle of* 40° *around the point* $(0, 1)$. *Describe the isometry* $\mathbf{R'R}$. *Is it the same as the isometry* $\mathbf{RR'}$?

We have taken care of translations and rotations. What about flips and glide reflections? Those are easy, too. If $\mathbf{F}(\alpha) = \bar{\alpha}$, then $\mathbf{F}$ is the reflection across the x-axis in the plane. Suppose we want to flip across the line through 0 and ρ, where as usual $\rho = \cos\theta + i\sin\theta$. If we first flip across the x-axis, that takes ρ to $\bar{\rho}$. Now if we rotate by an angle 2θ, that will bring ρ back to where it started. Since the origin is also unmoved, the line through 0 and ρ doesn't move. But then that must be the flip across that line. Our formula is $\mathbf{U}(z) = \rho^2\bar{z}$.

Problem

Find the general formula for a glide reflection. From the formula, read off the line across which the reflection is taken and the distance points move along that line.

Geometry of the Hyperbolic Plane

3.1 The Poincaré disc and Isometries of the Hyperbolic Plane

The Fifth Postulate of Euclid is equivalent to the statement that the sum of the angles in any triangle is equal to 180° (Hypothesis 5; see Section 2.1). From this we are able to deduce the possible regular and semiregular tilings of the plane (See Section 2.2). Now we are going to make the contrary assumption, that the sum of the angles in any triangle is less than 180°. (Recall Legendre's and Saccheri's Theorem 1.3.2, which says that either the sum of the angles in every triangle equals 180° or else the sum is always less than 180°.) This will lead to a very different conclusion about possible tilings. For example, four squares no longer fit together at their corners without leaving a gap. As we will see, however, it is possible for *five* squares to do so!

To begin, let us examine a model of the hyperbolic plane known as the Poincaré disc. Recall from Chapter 1 that a *model* for a postulate system is created by substituting specific objects for the undefined terms in the system in such a way that the postulates become true statements about the objects. In the example of Euclidean geometry known as Cartesian, or analytic, geometry, the word "point" meant an ordered pair (a, b) of real numbers and the word "line" meant the locus of points (a, b) satisfying the relation $as + bt + u = 0$ for some fixed triple (s, t, u) of real numbers with s and t not both equal to 0

What we are going to do is to describe a model for geometry in which all of the postulates except the parallel postulate are true and the parallel postulate fails. The model actually comes from Euclidean geometry, although we have to change our interpretation of some of the basic terms.

The hyperbolic plane $\mathcal{H}$ will consist of all points contained within the unit circle Ω with center at the origin. In terms of analytic geometry, $\mathcal{H}$ consists of all ordered pairs (a, b) of real numbers satisfying the inequality

$$a^2 + b^2 < 1.$$

Such an ordered pair is defined to be a **point**. A **line** is defined to be the locus of points lying inside Ω on a circle that meets the unit circle orthogonally (perpendicularly). To measure the **angle** between two (hyperbolic) lines, we draw the Euclidean straight lines tangent to the lines at the point of intersection and measure the angle between them. So for example, two hyperbolic lines are perpendicular if their tangents are perpendicular. We must also include in our definition of hyperbolic line the diameters of the circle Ω.

To understand this model, we need to investigate the geometry of circles and determine when two circles in the (Euclidean) plane are orthogonal to each other. The basic tool we will use is a geometric construction known as *inversion*.

Suppose Σ is a circle of radius k centered at a point O. If X is any point other than O, define its *inverse* X' with respect to Σ to be the point on the ray from O through X satisfying the equation

$$|OX| \times |OX'| = k^2.$$

Points inside Σ have inverses outside Σ. The inverse of X' is X. Points on the circle are their own inverses. The importance of inverses comes from the following lemma and its corollaries. To avoid confusion, note that the circle S in the lemma is not the circle through which we will be inverting points. That circle, Σ, will have center at the point O.

Lemma 3.1.1

If from a point O outside a circle S a line is drawn meeting the circle at X and X', then $|OX| \times |OX'| = k^2$, where k is the length of the tangent OP to S.

Proof

The key to this lemma is to see that triangles OXP and OPX' are similar triangles. Since they have one angle in common, we only need to check that $\angle XPO = \angle PX'O$. Let $\angle 1 = \angle QX'X = \angle QXX'$, $\angle 2 = \angle PX'O$, and $\angle 3 = \angle X'XP$. Then since $\Delta X'QP$ and ΔXQP are isosceles, $\angle QPX' = \angle 1 + \angle 2$, while

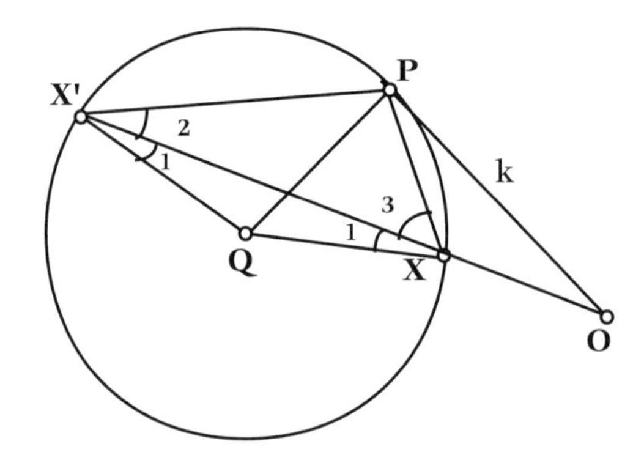

$\angle QPX = \angle 1 + \angle 3$. Add up the angles in $\Delta PXX'$ to see that $\angle 1 + \angle 2 + \angle 3 = 90°$. Now we use the fact that OP is tangent to the circle to get $\angle XPO = \angle 2$. Since triangles ΔOPX and $\Delta OX'P$ are similar, $\frac{|OX|}{|OP|} = \frac{|OP|}{|OX'|}$. ■

Corollary 3.1.2
If X and X' are inverses to each other with respect to the circle Σ, then any circle S passing through X and X' meets Σ orthogonally.

Proof
The tangent line OP from the center O of the circle Σ to the circle S satisfies the formula $|OP|^2 = |OX| \times |OX'| = k^2$, so $|OP| = k$. ■

Corollary 3.1.3
*If **F** is inversion through a circle Σ with center O and radius k, then **F** takes any circle orthogonal to Σ to itself.*

Proof
Let S be a circle that crosses Σ orthogonally at a point P. Then the radius OP of Σ is tangent to S. If a ray from O meets S at points X and Y, then by Lemma 3.1.1, $|OX||OY| = |OP|^2 = k^2$, so X and Y are inverses with respect to Σ. ■

Problem
Figure out how to construct X' given the circle Σ with center O and the point X, using only compass and straightedge. Then see if you can construct X' using only a compass. (This is easier when the point X is outside the circle.)

Proposition 3.1.4
*If **F** is inversion through a circle Σ with center O and radius k, then **F** takes any circle not passing through O to a circle not passing through O.*

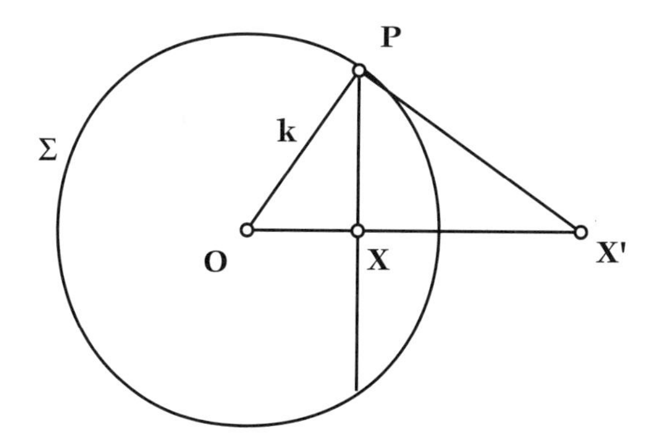

Proof

Suppose S is a circle with center C. For convenience let's assume that O is outside the circle. There is a similar argument when it is inside. The ray from O through C hits the circle at points A and B. Let P be any other point on the circle. Let $\mathbf{F}(A) = A'$, $\mathbf{F}(B) = B'$, and $\mathbf{F}(P) = P'$.

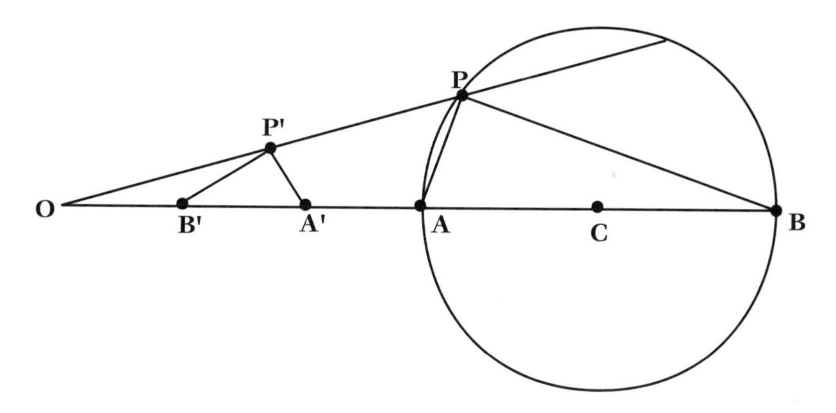

Now $|OP||OP'| = |OB||OB'|$, so $\frac{|OP|}{|OB|} = \frac{|OB'|}{|OP'|}$. This means that the triangles ΔOPB and $\Delta OB'P'$ are similar triangles, and $\angle OP'B' = \angle OBP$. By the same argument, $\angle OP'A' = \angle OAP$. Furthermore, $\angle OAP = \angle OBP + \angle APB$ (because an exterior angle is the sum of the other two interior angles—we are still doing Euclidean geometry here!) So $\angle APB = \angle A'B'P'$.

But AB is a diameter of a circle and P is on that circle, so $\angle APB = 90°$. So if we draw the circle with diameter $A'B'$, then by the same principle, P' must be on that circle. The circle with diameter AB is therefore taken to the circle with diameter $A'B'$. ■

Problem

Show that the inverse of a circle through O is a straight line not passing through O.

In the proof of Proposition 3.1.4 we saw that if three points A, P, and B were taken to points A', P', and B', the corresponding angles $\angle APB$ and $\angle B'P'A'$ are equal. Using this observation it is possible to prove:

Proposition 3.1.5

Inversion through a circle Σ takes curves crossing at an angle α to curves crossing at the same angle α.

Now we will apply Corollary 3.1.2 to the circle Ω. If X is any point inside Ω, then there is a corresponding point X' outside. If X happens to be the center O, the construction does not apply, and we can think of the other point as being "at infinity." Let's worry about this later.

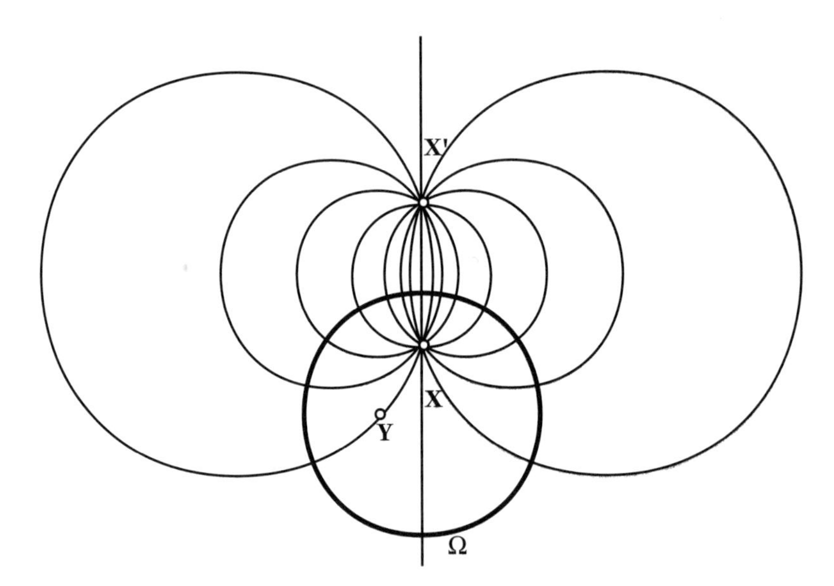

If X and Y are two points in Ω, there is a unique circle passing through X, Y, and X'. (This is a fact from Euclidean geometry.) By Corollary 3.1.2, this circle will be orthogonal to Ω, so the part inside Ω is a hyperbolic straight line (which we will call an *h-line* for short). This proves that two points are uniquely joined by an h-line. The collection of h-lines through X is the set of circles through X and X'. (This is known as a *coaxial system of circles*.)

In the special case where X is the center O of the circle, this construction is replaced by the easier one of taking all the ordinary straight lines through O. We can imagine these lines as all passing through a point O' at infinity. Using the construction known as *stereographic projection*, there is another way to visualize this.

Imagine the Euclidean plane **P** as the x–y plane in three–dimensional space, and let **S** be the sphere of radius 1 centered at the origin. We'll call the north pole of this sphere N; it has coordinates $(0, 0, 1)$. If A is any other point on **S**, then the line through N and A intersects **P** at the point $\sigma(A)$. σ is called stereographic projection. This construction goes back a couple of thousand years. If the sphere is thought of as the surface of the earth, then σ sets up a correspondence between points on the earth's surface and points in a flat plane; in other words, a *map* of the world. Such a map was known to Ptolemy, who used it in his book *Geographia*.

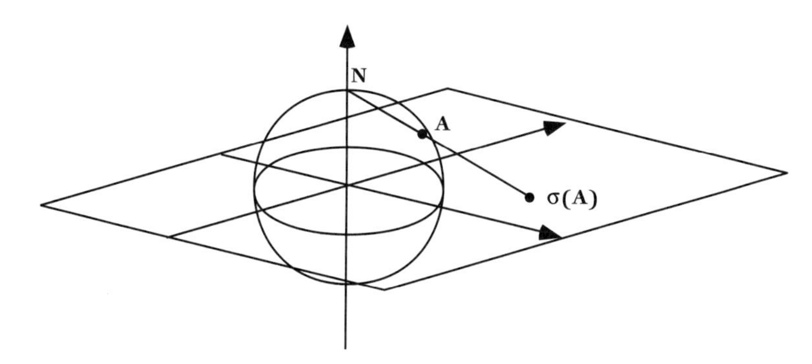

Stereographic projection has a number of useful properties, of which two are important for us right now. The first is that it takes circles to circles and straight lines. Circles on **S** that do not pass through N are taken to circles in **P**; circles passing through N are taken to straight lines. Thinking of this in the opposite direction, straight lines in **P** correspond to circles through the north pole of **S**. We can think of N as "the point at infinity" where straight lines all meet.

The second property of stereographic projection is that if two circles meet at an angle θ on **S**, then their images under stereographic projection also meet at angle θ. The map σ is said to be *conformal.*

Problem

An exercise in analytic geometry: If $\sigma(a, b, c) = (x, y, 0)$, *find the equations for* x *and* y *in terms of* a, b, *and* c. *If a straight line in the plane is given by the equation* $sx + ty + u = 0$, *determine the corresponding relationship between* a, b, *and* c *and verify that this describes a circle passing through the north pole.*

Stereographic projection takes the southern hemisphere of **S** to the points in $\mathcal{H}$, the points inside the unit circle Ω, and the equator corresponds to Ω. A circle on the sphere crosses the equator orthogonally precisely when its image under stereographic projection crosses Ω orthogonally. Now, how do we get such a circle on the sphere? It's simple! Just take a vertical plane that slices through the sphere. The intersection of a plane with a sphere is a circle, and vertical planes cut the equator orthogonally. The circles in the plane passing through a point X correspond to circles passing through a point A on the sphere, where $X = \sigma(A)$. So take the vertical line through A and look at all planes containing that line. These planes slice through the sphere along circles through A that cross the equator orthogonally. In this picture, the circle that also passes through the north pole corresponds to the radial straight line through X.

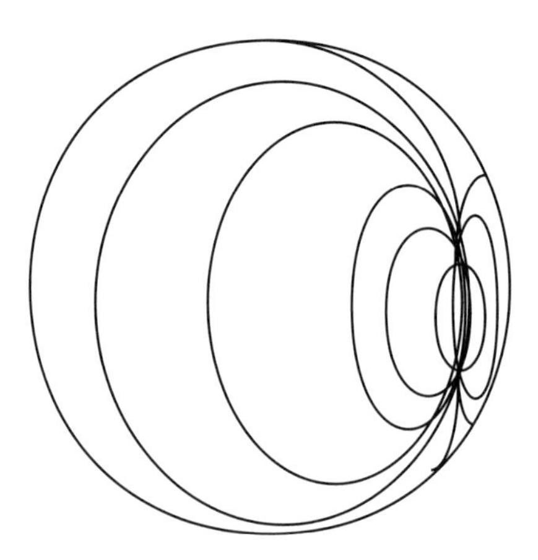

The vertical line through A meets the sphere at a second point A'. If $\sigma(A) = X$, then $\sigma(A') = X'$, where X' is the inverse of X with respect to the unit circle. So the relationship of inverse points on the sphere is quite simple. In the case where A is the south pole of the sphere, it becomes natural to think of the north pole as its inverse. The circles through the poles orthogonal to the equator are the meridians of the sphere, which correspond to the straight lines through the origin in the plane. Now we see that those straight lines are just like arcs of circles that hit Ω orthogonally.

So far, we have verified that in the Poincaré disc model of hyperbolic geometry two points determine a line. To verify that Euclid's third and fourth postulates hold in this model, we need to have a notion of congruence. Suppose ℓ is an h-line in $\mathcal{H}$. That is, ℓ is the part of a circle S crossing Ω orthogonally. Define $\mathbf{F}$ by the rule $\mathbf{F}(X) = X'$ if X' is the inverse of X with respect to the circle S. $\mathbf{F}$ will be the *flip*, or *reflection*, through ℓ in $\mathcal{H}$. In the special case where ℓ is part of a straight line, $\mathbf{F}$ is just the ordinary reflection through that line.

Because Ω is orthogonal to S, points on Ω have inverses that are on Ω. Therefore, $\mathbf{F}(X)$ is in $\mathcal{H}$ whenever X is in $\mathcal{H}$. We will declare $\mathbf{F}$ to be an isometry of the hyperbolic plane, and therefore any composition of these flips will be an isometry. This should take care of all possible isometries, because of a fundamental theorem of geometry:

Theorem 3.1.6 *Any isometry of the plane is a composition of at most three reflections.*

Proof

Let AB be a line segment, and let $A'B'$ be a congruent segment. We saw in Chapter 2 that there are exactly two isometries that carry A to A' and B to B' (Pasch's axiom). Let's see how to find two isometries using only reflections. If ℓ_1 is the perpendicular bisector of the segment AA', then the reflection $\mathbf{F}_1$ across ℓ_1 will take A to A'. Suppose $\mathbf{F}_1(B) = B''$. If $B'' = B'$, then we hit the jackpot and can stop here. Otherwise, let ℓ_2 be the line bisecting the angle $\angle BA'B''$. Then the flip $\mathbf{F}_2$ will take B'' to B', and the isometry we want is $\mathbf{F}_2\mathbf{F}_1$. We have found one isometry; the other is gotten by following the first by the flip across the line through A' and B'. ■

This theorem is a classical theorem of Euclidean geometry, whose proof does not use the parallel postulate. Consequently it is also a true theorem of hyperbolic geometry. In general, geometrical theorems that do not depend on the parallel postulates are theorems in what is called *absolute geometry*.

Problem

Show that a rotation about a point X through an angle θ can be accomplished by two flips. Show that a translation along a line ℓ can be achieved by two flips.

That our definition of isometry makes sense requires a bit of work. By Proposition 3.1.4, $\mathbf{F}$ takes circles to circles. By Proposition 3.1.5, a circle orthogonal to Ω will be taken to a circle orthogonal to Ω. So h-lines are taken to h-lines and angles are preserved. We need to know also that if A and B are two points in $\mathcal{H}$, then there are exactly two isometries that fix both of them. In Section 3.3 we will be able to prove this algebraically.

If λ is an h-line and X is a point not on λ, then there are infinitely many h-lines through X that do not meet λ. Among them are two h-lines μ and ν that pass through the same points of Ω. These are sometimes called "parallel" to λ, while the other lines are called "hyperparallel." The line λ and the two parallels through X together form a sort of triangle with two of its three vertices on Ω. A point on the circle Ω is called an *ideal point*. An ideal point is not a point in $\mathcal{H}$, but rather a "point at infinity." If Y is

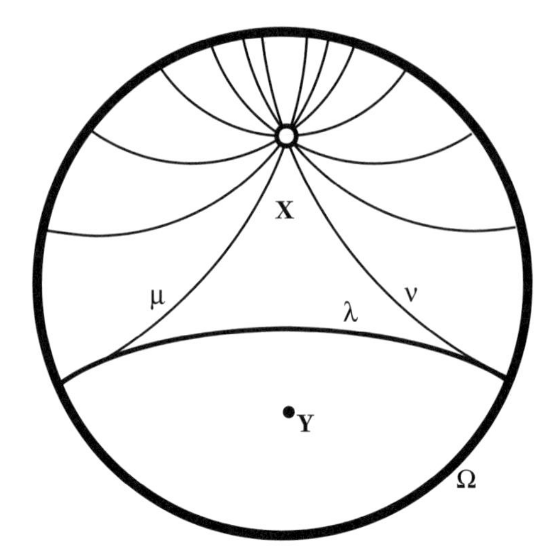

any point on the opposite side of λ from X, then it is not possible to draw an h-line through Y that intersects both μ and ν.

This explains the flaw in Legendre's proof of Theorem 1.3.4 of Chapter 1. In that proof we had a triangle ABC and a point D below the segment BC. We assumed that there was a line through D that met two rays emanating from A. (The picture in the proof showed the line EF already drawn.) The existence of EF used the parallel postulate!

Ideal points are very helpful in understanding the geometry of the hyperbolic plane. Given two ideal points $\mathcal{P}$ and $\mathcal{Q}$, there is a unique h-line λ that "connects" them. If $\mathbf{T}$ is a (hyperbolic) translation along λ, then as a set, the line λ is carried to itself. It is *invariant* under $\mathbf{T}$. Since our isometries are represented by transformations of the Poincaré disc, we can think of them as also defined on the ideal points. Ideal points are always taken to ideal points by any isometry of $\mathcal{H}$. It follows that the isometry fixes the points $\mathcal{P}$ and $\mathcal{Q}$. (Where else could they go?)

These turn out to be the only two ideal points that are fixed by $\mathbf{T}$. But this means that no other line is invariant under $\mathbf{T}$. Unlike the Euclidean case, where translation along a line is also translation along all parallel lines, a translation in hyperbolic geometry only slides one line along itself. Instead, points at a fixed distance h from the line λ are carried to points the same distance from λ. (We have not actually defined distance. This can be done, however, and then our isometries will preserve distances.) The set of such points is called an *equidistant curve*. In the disc, this curve is described by a circle arc passing through the points $\mathcal{P}$ and $\mathcal{Q}$.

How can we see that this is so? We know that the isometry $\mathbf{T}$ fixes $\mathcal{P}$ and $\mathcal{Q}$. We know that circles go to circles, so every circle through $\mathcal{P}$ and

$\mathcal{Q}$ goes to some circle through $\mathcal{P}$ and $\mathcal{Q}$. Since **T** preserves angles, each circle through $\mathcal{P}$ and $\mathcal{Q}$ must in fact be taken to itself.

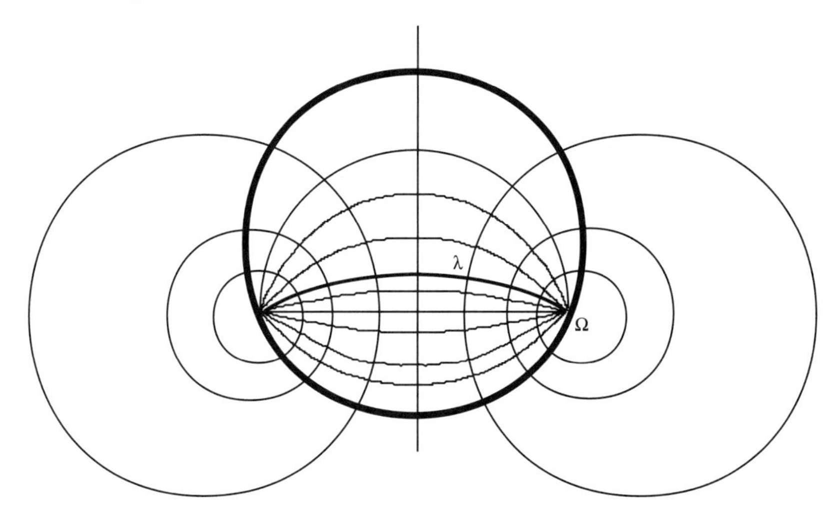

The h-lines orthogonal to λ correspond to circles that are orthogonal to all circles through $\mathcal{P}$ and $\mathcal{Q}$. Translation carries one such line to another. In the picture above, the light circles correspond to h-lines, while the horizontal curves are loci of points a fixed distance from the h-line λ. As the curves approach an ideal point, they appear to our Euclidean eyes to be getting closer together, but they are not. All h-line segments joining two equidistant curves are congruent to each other. As we move an object closer and closer to Ω while keeping its shape the same from the point of view of hyperbolic geometry, it appears to our Euclidean eyes to be shrinking in size. In the next section, we will be tiling the hyperbolic plane with congruent polygons. In the Poincaré disc, the polygons will appear to get small as they get close to the boundary circle.

Problem

Suppose λ and μ are two h-lines that meet at an ideal point $\mathcal{P}$. Let $\mathbf{F}_1$ be the reflection across λ and $\mathbf{F}_2$ the reflection across μ. Show that $\mathbf{F}_2\mathbf{F}_1(X) \neq X$ for all points in $\mathcal{H}$. Show that $\mathbf{F}_2\mathbf{F}_1(\mathcal{Q}) \neq \mathcal{Q}$ for any ideal point other than $\mathcal{P}$. Conclude that this isometry is not a translation, rotation, or glide reflection.

3.2 Tessellations of the Hyperbolic Plane

A regular polygon in $\mathcal{H}$ is, as in the Euclidean case, a polygon whose edges are all congruent and whose vertex angles are all congruent. Here is a simple way to make one. Let's suppose we want to make a regular hexagon. Draw six rays from the center O of Ω at 60° angles extending to Ω. Draw a circle centered at O; it meets the rays at six points. Now connect consecutive points with h-lines instead of regular lines. Since the ordinary Euclidean rotation around O is an isometry of $\mathcal{H}$, the polygon is regular.

Note that we can construct a regular four-sided polygon by this method. But we probably should not call it a "square," since the angles at the four corners are not right angles. This creates a bit of confusion about the idea of "area" of a hyperbolic figure, since there is no "unit square" floating around. Actually, area is not all that simple a concept even in Euclidean geometry. There are three simple properties of area that are easy to describe. The first is that if one polygon is inside another, then the inner polygon has smaller area than the outer one. The second property is that area is *additive*; this means that if a polygon is cut up into smaller polygons, then the area of the large polygon is equal to the sum of the areas of the smaller polygons. The third is that if two polygons are congruent, then they have the same area. In Euclidean geometry, we would then choose a unit square and use these principles to determine the areas of polygons. If we get into the question of objects with curved sides, the subject gets quite a bit more complicated, so we will stay with polygons.

In Section 1.3 we defined the *defect* of a triangle to be the difference between 180° and the sum of the angles in a triangle, and we saw that the defect is also *additive*. That is, if a triangle is cut up into smaller triangles, then the defect of the triangle equals the sum of the defects of the smaller triangles. Using this fact it is not too difficult to show:

Lemma 3.2.1
The area of a triangle is proportional to its defect. There is an upper bound to the area of a triangle in $\mathcal{H}$.

This is a remarkable fact! There is no comparable phenomenon in Euclidean geometry. In fact, we may *define* the area of a triangle to equal its defect; then every triangle has area smaller than 180. (Actually, it is customary to use radians rather than degrees, so that every triangle has area less than π.)

Turning to our hexagons again, if we connect the ideal points at the ends of the six rays from the origin by h-lines, we get an "ideal hexagon." Since h-lines are parts of circles orthogonal to Ω, the adjacent sides of this polygon are actually tangent; in effect, the vertex angles are 0. So as we construct regular hexagons with vertices chosen farther and farther away from the origin, the vertex angles are decreasing, and they can be made as close to 0 as we like. In fact, if θ is any angle less than 120°, there is exactly one way to construct a regular pentagon with interior angles all equal to θ.

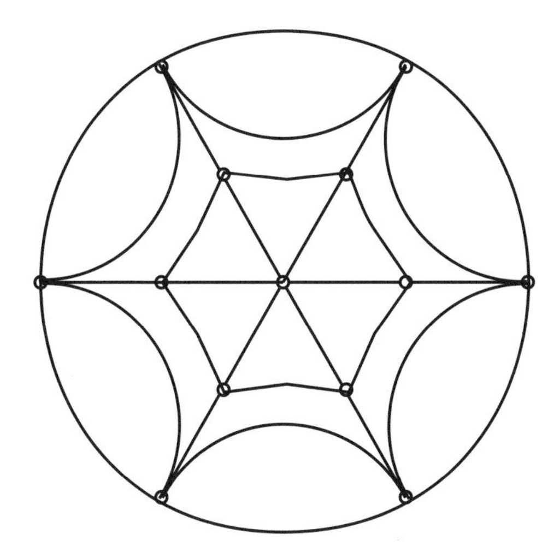

We have built our hexagon centered at the origin. By using an isometry, we can move it to any place we like. With our "Euclidean" eyes, however, the polygon will appear to shrink as we move it farther away from the origin. For example, we could start with a hexagon whose interior angles are exactly 90°. By reflecting the hexagon across each of its six sides, we get six more hexagons, each apparently smaller than the original but actually congruent to it in $\mathcal{H}$. Since there is 90° left outside of this figure at each of the six vertices, we can just squeeze one more hexagon in. This is the beginning of the process of building the regular tiling (6, 6, 6, 6) of $\mathcal{H}$ by hexagons with four at each vertex.

Of course, we could use pentagons instead of hexagons. Now the interior angles must be smaller than 108°; we can find a right-angled pentagon and use it to tile the plane. Below is a picture of the regular tiling (5, 5, 5, 5).

By making the pentagon bigger, we can reduce the vertex angles to 72°. This gives the (5, 5, 5, 5, 5) regular tiling of the plane.

We can also construct semiregular tilings. For example, we might try to put two regular hexagons and a regular octagon together at a vertex. But how do we know we can do this? In the Euclidean case, if we wanted to put two octagons and a square together, we just chose the square with the same edge length as the octagon and then put them together. The angles automatically added up. But in $\mathcal{H}$ life is more interesting. We can build an octagon and two hexagons with the same edge length, but the angles might not add up right to put them together.

Suppose we start by building the three polygons very large. Then the sum of the angles around the vertex can be made smaller than 360° without any difficulty. Now start shrinking the polygons, keeping the

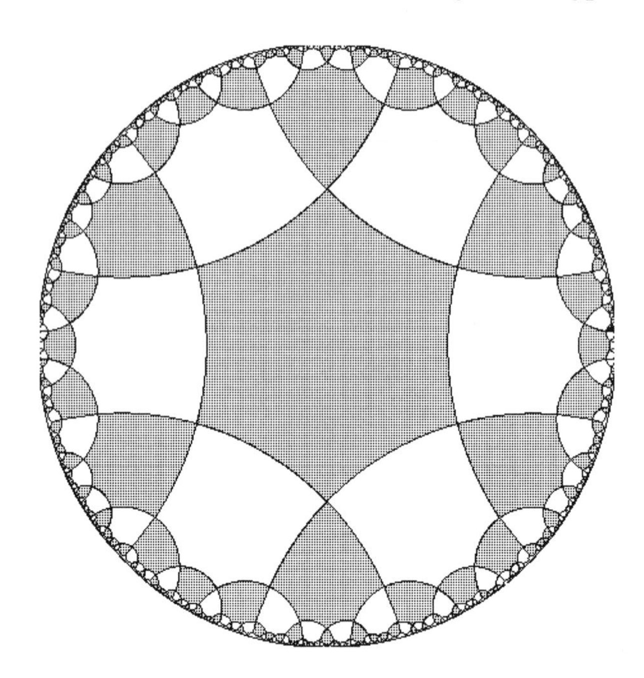

edge lengths equal to each other. The angles will keep growing. In fact, the smaller we make the polygons, the closer the angles are to the Euclidean angles (because the defect is proportional to area by Lemma 3.2.1). So there is some intermediate size at which they will just fit together. This is the size we will use for the $(6, 6, 8)$ semiregular tessellation of $\mathcal{H}$.

This is all very well and good, but how do we know that this process always works? In the Euclidean case, we were able to put two pentagons and a hexagon together at a vertex, but the pattern could not be extended to a tiling of the plane. What assurance do we have that there really is a $(6, 6, 8)$ tiling?

One way to approach this problem is to look at the *dual* tiling. Suppose we have actually constructed the $(6, 6, 8)$ tiling. As we did in Chapter 2, construct a tiling by taking as vertices the centers of the polygons and as edges the h-line segments joining two adjacent polygons. This will divide $\mathcal{H}$ into triangles, all of them identical, one for each vertex of the original tiling. The angles at the vertices will be exactly $60°$, $60°$, and $45°$. (This follows from the fact that we cut up each polygon into identical pieces, and the central angles add up to $360°$.) Any two adjacent triangles are mirror images of each other.

Now let's turn the problem around. Start with the triangle with vertex angles $60°, 60°$, and $45°$. If we can generate a tiling of $\mathcal{H}$ by repeatedly reflecting the triangle across its sides, then its dual tiling will be the $(6, 6, 8)$ tiling we seek.

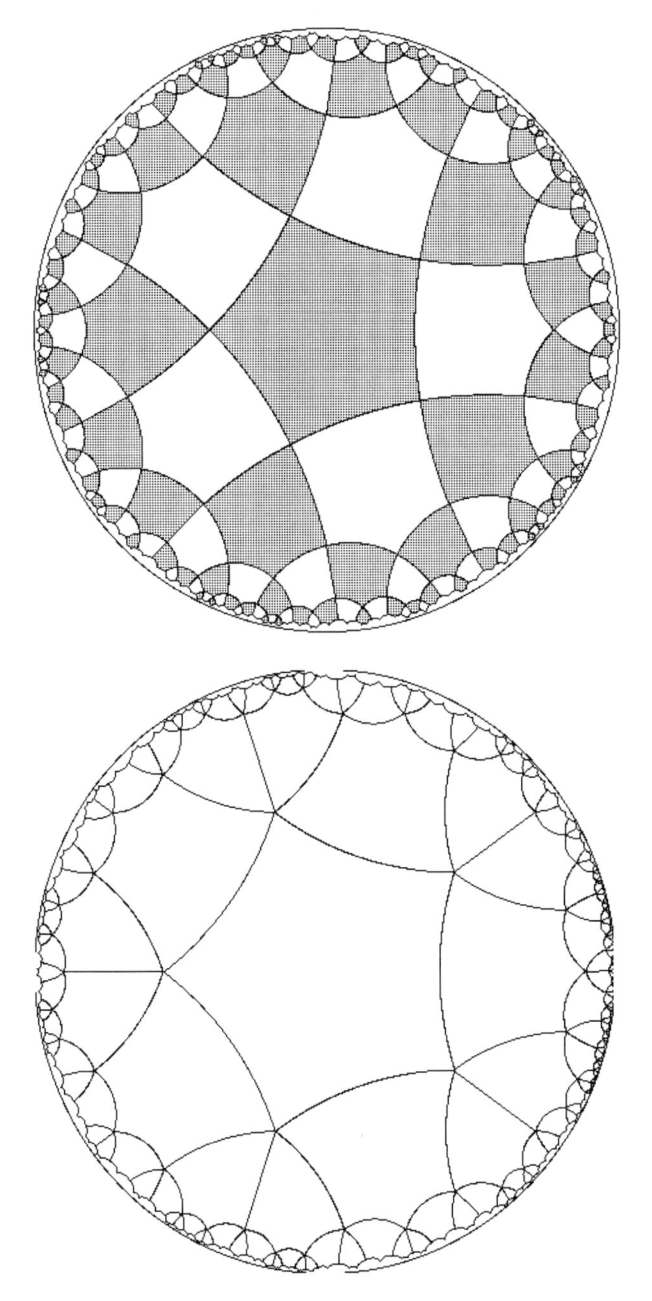

This turns out always to be possible. Namely, if l, m and n are integers that satisfy the equation

$$\frac{1}{l} + \frac{1}{m} + \frac{1}{n} < 1,$$

then the triangle T whose angles are $\frac{180}{l}$, $\frac{180}{m}$, and $\frac{180}{n}$ degrees tiles the hyperbolic plane by repeated reflection across its sides. The dual tiling to this tiling by triangles is the $(2l, 2m, 2n)$ tiling.

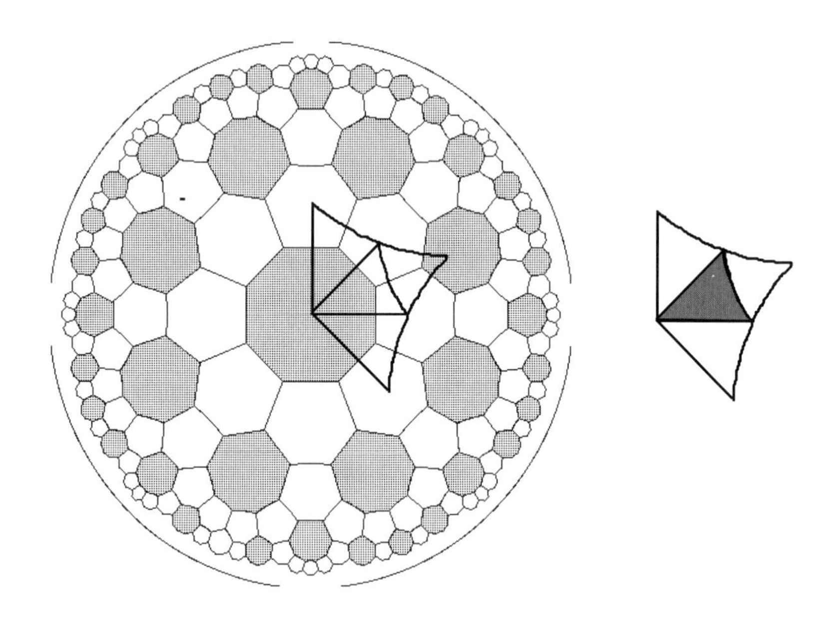

Let $\mathbf{F}_1$, $\mathbf{F}_2$, an $\mathbf{F}_3$ be the reflections across the three sides of T. If we alternately flip across two of the sides, the vertex where they meet does not move, and the result is a rotation around that vertex. We get the equation $(\mathbf{F}_2\mathbf{F}_1)^n = \mathbf{I}$, where $\mathbf{I}$ is the identity transformation (See Section 2.1 for information about transformations.) Likewise, $(\mathbf{F}_3\mathbf{F}_2)^l = \mathbf{I}$, and $(\mathbf{F}_1\mathbf{F}_3)^m = \mathbf{I}$. If we color T black, color the triangles adjacent to T white, and then follow the rule that whenever we flip a triangle it reverses color, then the triangles will be neatly divided into two sets, with adjacent triangles never having the same color.

Problem

Modify the construction above to account for the fact that there is also a $(7, 6, 6)$ *tiling of the hyperbolic plane.*

A proof that repeated flipping of a triangle across its sides does not lead to inconsistencies is actually a bit too advanced for this book. It requires some *topology*, specifically, the theory of covering spaces. An intuitive explanation is possible, however. It is easier to describe in the Euclidean case, but the non-Euclidean argument would be equally fine (if we happened to be non-Euclidean people). Imagine a room with three walls that are covered with mirrors and a floor that is in the shape of a

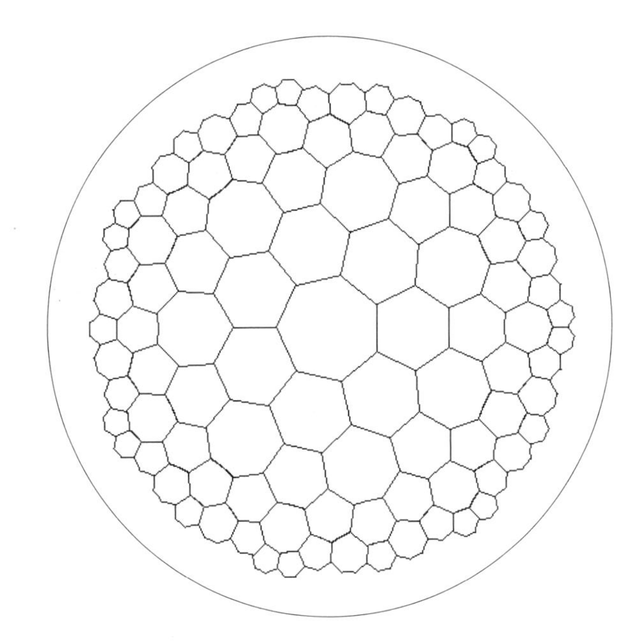

triangle with each angle evenly dividing 180°. (Unfortunately, there is rather a narrow selection, since we are Euclidean!) As we look at a wall we seem to see a vista (with lots of images of ourselves seen from different angles). Because of the angle condition, whenever we see the edge of the room, everything around it fits together perfectly. The result is that we see a plane covered with trianglular flooring, one triangle for each image of the room. That is our tiling.

Problem

Suppose we were in a room with three triangular walls meeting at a point and a triangular floor (in other words, we were inside a tetrahedron). Suppose all the walls and the floor had mirrors. Would it be possible for the view to seem like ordinary (Euclidean) space in all directions? In other words, is it possible to fill up space with reflected copies of a tetrahedron? If so, how?

M.C. Escher Maurits Cornelis Escher (1898–1972) studied graphic arts at the School of Architecture and Decorative Arts, in Haarlem. From his student days on, he was fascinated with repetetive design patterns, and throughout his life he produced marvelous designs in woodcut and wood engraving using the geometry of planar tessellations. A typical woodcut features a repeating pattern made by taking one of the possible symmetry groups and replacing the geometric figures by birds, fish, or other real or imaginary creatures.

According to Doris Schattschneider's book *Visions of Symmetry* ([30], P. 251), Escher was intrigued with the idea of artistically representing infinity by a repeated pattern that diminished in size. One of his designs, *Whirlpools* (November 1957), depicts two spiraling streams of fish swimming in opposite directions along spirals that emerge from one vortex and disappear into another. The fish diminish in size as they approach the center of each vortex. Escher struggled for years with the problem of producing a design in which instead the fish would recede endlessly toward the boundary of an enclosed region while decreasing in size. Then in 1958 he found a mathematical article by H.S.M. Coxeter (*Crystal Symmetry and its Generalizations*, in "a Symposium on Symmetry," Transactions of the Royal Society of Canada **51** (1957), 1–13) which pictured a hyperbolic tessellation of the plane. This led to correspondence between Escher and Coxeter, and the magnificent woodcut *Circle Limit III*, displaying yellow, green, blue, brown, and black fish inside a disc, whose backbones form the edges of the semiregular tiling $(4, 3, 4, 3, 4, 3)$ of $\mathcal{H}$.

Doris Schattschneider's book [30] is a superb source for information about Escher and his work. The geometry of the hyperbolic plane is discussed in [4], [7], [35], and [36].

3.3 Complex numbers, Möbius Transformations, and Geometry

Just as complex numbers are useful in describing the isometries of the Euclidean plane, so are they useful in describing the isometries of the

hyperbolic plane. There is a simple reason for this. We will find out that inversion through a circle can be described simply using complex arithmetic. Since every transformation of $\mathcal{H}$ is a composition of such isometries, the formulas for isometries can be derived by understanding inversions. It turns out that every isometry can be written by a formula of the form

$$\mathbf{U}(z) = \frac{\alpha z + \beta}{\gamma z + \delta}$$

or of the form

$$\mathbf{U}(z) = \frac{\alpha \bar{z} + \beta}{\gamma \bar{z} + \delta},$$

where the quantity $\Delta = \alpha\delta - \beta\gamma$ is assumed to be nonzero. (If $\Delta = 0$, then $\mathbf{U}$ is constant.) Transformations of the complex plane given by the first type of formula are called *Möbius transformations,* while those of the second type are called *conjugate Möbius transformations*. Not all of these transformations are actually isometries of $\mathcal{H}$; we must single out those transformations that take points inside the unit circle Ω to points inside Ω.

Let's begin with a simple example. Suppose we want to invert points in the plane with respect to the unit circle. That means we must take every point X to a point X' on the ray through O and X in such a way that $|OX||OX'| = 1$. If α is the complex number corresponding to the point

X, then any point on the ray through O and X can be represented as a complex number $r\alpha$, where r is a positive real number. Since $|OX| = |\alpha|$, the equation for inverses is $|\alpha||r\alpha| = 1$, or $r = \frac{1}{|\alpha|^2}$. Now using the fact that $|\alpha|^2 = \alpha\bar{\alpha}$ and plugging in for r in the expression for X', we get

Proposition 3.3.1

The inversion through the unit circle is given by the formula

$$\mathbf{F}_1(\alpha) = \frac{1}{\bar{\alpha}}$$

If we want to invert through the circle of radius k centered at the origin, we can modify the argument above and quickly see that the appropriate formula is

$$\mathbf{F}_k(\alpha) = \frac{k^2}{\bar{\alpha}}$$

Now, what about inversion through a circle not centered at the origin? Here is a simple strategy for finding the formula. Suppose $\mathbf{T}_X$ is the translation that takes the origin to the point X. $\mathbf{T}_X^{-1}$ takes X back to the origin. To accomplish the inversion through the circle of radius k centered at X, we shift X to the origin, perform the inversion through the circle of radius k centered at the origin, then shift back to X. If α is the complex number that corresponds to the point X, then the formula for translation is $\mathbf{T}_X(z) = z + \alpha$. The inversion is then given by

$$\begin{aligned}
\mathbf{F}(z) &= \mathbf{T}_X \mathbf{F}_k \mathbf{T}_X^{-1}(z) \\
&= \mathbf{T}_X \mathbf{F}_k(z - \alpha) \\
&= \mathbf{T}_X\left(\frac{k^2}{\bar{z} - \bar{\alpha}}\right) \\
&= \left(\frac{k^2}{\bar{z} - \bar{\alpha}}\right) + \alpha \\
&= \left(\frac{\alpha\bar{z} + (k^2 - |\alpha|^2)}{\bar{z} - \bar{\alpha}}\right).
\end{aligned}$$

Every inversion through a circle is a conjugate Möbius transformation. We are also interested in reflection across a line through the origin, since that also corresponds to a hyperbolic reflection. Recall from Chapter 2 that the formula for such a reflection is $\mathbf{F}(z) = \rho^2\bar{z}$. This too is a conjugate Möbius transformation (with $\gamma = 0$).

Suppose **U** and **U'** are conjugate Möbius transformations. We need to be able to compute their composition. This is just a brute force calculation:

$$
\begin{aligned}
\mathbf{U'U}(z) &= \mathbf{U'}\left(\frac{\alpha\bar{z}+\beta}{\gamma\bar{z}+\delta}\right) \\
&= \frac{\alpha'\overline{\frac{\alpha\bar{z}+\beta}{\gamma\bar{z}+\delta}}+\beta'}{\gamma'\overline{\frac{\alpha\bar{z}+\beta}{\gamma\bar{z}+\delta}}+\delta'} \\
&= \frac{\alpha'\overline{\alpha\bar{z}+\beta}+\beta'\overline{\gamma\bar{z}+\delta}}{\gamma'\overline{\alpha\bar{z}+\beta}+\delta'\overline{\gamma\bar{z}+\delta}} \\
&= \frac{(\alpha'\bar{\alpha}+\beta'\bar{\gamma})z+(\alpha'\bar{\beta}+\beta'\bar{\delta})}{(\gamma'\bar{\alpha}+\delta'\bar{\gamma})z+(\gamma'\bar{\beta}+\delta'\bar{\delta})}.
\end{aligned}
$$

So the composition of two conjugate Möbius transformations is a Möbius transformation. A similar computation shows that a composition of a Möbius transformation and a conjugate Möbius transformation is a conjugate Möbius transformation. Since reflections in hyperbolic geometry can be represented by inversions through a circle, and since by Theorem 3.1.6 every isometry is a composition of such reflections, this gives us:

Theorem 3.3.2 *Any isometry of the hyperbolic plane can be represented by a Möbius or congugate Möbius transformation. In particular, translations and rotations can be represented by Möbius transformations, while reflections and glide reflections can be represented by conjugate Möbius transformations.*

Any conjugate Möbius transformation $\mathbf{U}(z)$ can be written as a composition of simple transformations. If $\gamma \neq 0$, then

$$
\mathbf{U}(z) = \frac{\alpha\bar{z}+\beta}{\gamma\bar{z}+\delta} = \frac{\alpha}{\gamma} - \frac{\alpha\delta-\beta\gamma}{\gamma(\gamma\bar{z}+\delta)} = \frac{\alpha}{\gamma} - \frac{\alpha\delta-\beta\gamma}{\gamma^2}\frac{1}{(\bar{z}+\frac{\delta}{\gamma})}.
$$

This formula shows that every conjugate Möbius transformation can be written as a composition of a translation $z \longmapsto z + (\overline{\frac{\delta}{\gamma}})$; a circle inversion $z \longmapsto \frac{k^2}{\bar{z}}$ where $k^2 = \left|\frac{\Delta}{\gamma^2}\right|$; the rotation $z \longmapsto \rho^2 z$, where $\rho^2 k^2 = -\frac{\Delta}{\gamma^2}$; and the translation $z \longmapsto z + \frac{\alpha}{\gamma}$. Of course, if $\gamma = 0$, the formula does not apply. In that case, $\mathbf{U}$ is a composition of reflection across a line, a dilation, and a translation.

Problem

Decompose Möbius transformations into simple transformations as above. Using this decomposition, conclude that every Möbius or conjugate Möbius transformation takes lines and circles to lines and circles and preserves angles.

The Möbius transformations form a group $\mathcal{M}$ under composition. Using the stereographic projection of the sphere onto the plane (Section 3.1), we can view $\mathcal{M}$ as transformations of the sphere. Remember that the north pole of the sphere corresponds to the "point at infinity" in the plane. If we then interpret "infinity" as the reciprocal of 0, the Möbius transformation can be thought of as defined on this "extended complex plane" $\mathbf{C}^*$, which is known as the *inversive* or *conformal plane* by geometers ([7], p. 84; [29], p. 75) and the *Riemann sphere* in complex analysis. The obvious advantage of this is that then the transformation $\mathbf{U}(z) = \frac{\alpha z+\beta}{\gamma z+\delta}$ is defined for every complex number z, including "∞." We rewrite

$$\mathbf{U}(z) = \frac{\alpha + \frac{\beta}{z}}{\gamma + \frac{\delta}{z}}$$

and see that $\mathbf{U}(\infty) = \frac{\alpha}{\gamma}$, while $\mathbf{U}(-\frac{\delta}{\gamma}) = \infty$.

Thought of as a transformation of the sphere, $\mathbf{U}$ is a *conformal* map, a function that preserves angles. It is a deep theorem of complex analysis that Möbius transformations are the *only* conformal transformations of the sphere that take distinct points to distinct points. $\mathbf{U}$ also takes circles to circles.

Problem

Let $\mathbf{U}(z) = \frac{\alpha z+\beta}{\gamma z+\delta}$. *If* U *is not the identity transformation, what are the possibilities for the number of points* z *for which* $\mathbf{U}(z) = z$*? (Include* ∞ *as a possible point.) Investigate the same question for conjugate Möbius transformations.*

Problem

The cross-ratio *of four complex numbers* z_1, z_2, z_3, *and* z_4 *is the quantity*

$$(z_1, z_2; z_3, z_4) = \frac{(z_1 - z_3)(z_2 - z_4)}{(z_2 - z_3)(z_1 - z_4)}.$$

If $\mathbf{U}$ *is a Möbius transformation, show that*

$$(\mathbf{U}(z_1), \mathbf{U}(z_2); \mathbf{U}(z_3), \mathbf{U}(z_4)) = (z_1, z_2; z_3, z_4).$$

If z_1, z_2, z_3 *are three distinct complex numbers, and if* w_1, w_2, *and* w_3 *are distinct complex numbers, verify that the equation* $(w, w_2; w_3, w_4) = (z, z_2; z_3, z_4)$ *defines a Möbius transformation* $w = \mathbf{U}(z)$ *for which* $w_i = \mathbf{U}(z_i)$.

Among the Möbius transformations, those that take the interior of the unit disc to itself form a subgroup $\mathcal{G}$. For each transformation $\mathbf{U}$ in $\mathcal{G}$, there is a conjugate Möbius transformation $\bar{U}$ defined by $\bar{U}(z) = \overline{U(z)}$. There is also a Möbius transformation $\mathbf{U}^*$ defined by $\mathbf{U}^*(z) = \overline{\mathbf{U}(\bar{z})}$. From the properties of conjugation, $\overline{UV} = \overline{UV}$ and $\bar{\mathbf{U}}\bar{V} = \mathbf{U}^* V$. The transformations in $\mathcal{G}$ and their conjugates form the isometry group $\mathcal{I}(\mathcal{H})$. To understand this group, we really only need to understand $\mathcal{G}$. Let's figure out which Möbius transformations belong to $\mathcal{G}$.

Problem

Verify the formulas $\overline{U}V = \overline{UV}$ *and* $\bar{\mathbf{U}}\bar{V} = U^*V$. *When is* $U^* = U$?

If $\mathbf{U}$ is in $\mathcal{G}$, then whenever $z\bar{z} = 1$, we must also have $\mathbf{U}(z)\overline{U(z)} = 1$. Clearing the denominator from this equation, we get

$$(\alpha z + \beta)\overline{(\alpha z + \beta)} - (\gamma z + \delta)\overline{(\gamma z + \delta)} = 0.$$

Multiplying this out and replacing $z\bar{z}$ by 1, this becomes

$$(\alpha\bar{\beta} - \gamma\bar{\delta})z + (\bar{\alpha}\beta - \bar{\gamma}\delta)\bar{z} + (\alpha\bar{\alpha} - \gamma\bar{\gamma}) + (\beta\bar{\beta} - \delta\bar{\delta}) = 0.$$

The two quantities $(\alpha\bar{\beta} - \gamma\bar{\delta})z$ and $(\bar{\alpha}\beta - \bar{\gamma}\delta)\bar{z}$ are complex conjugates of each other, so they add up to $2\mathbf{Re}(\alpha\bar{\beta} - \gamma\bar{\delta})z)$ (twice the real part). This is supposed to be constant, but z can be any number for which $|z| = 1$. So the coefficient of z has to be 0. (If this is not clear, try $z = 1$, $z = -1$, $z = i$, and $z = -i$.) So these equations are not so bad after all. They become

$$\alpha\bar{\beta} - \gamma\bar{\delta} = 0, \tag{3.3.1}$$

$$|\alpha|^2 - |\gamma|^2 = |\delta|^2 - |\beta|^2. \tag{3.3.2}$$

If $\alpha = 0$, we are quickly led to $\mathbf{U}(z) = \frac{\kappa}{z}$, with $|\kappa| = 1$. This does take the unit circle to itself, but it switches the inside and the outside. So let's choose a complex number κ by the equation $\bar{\delta} = \bar{\kappa}\alpha$. Then equation 3.3.1 tells us that $\bar{\beta} = \bar{\kappa}\gamma$. Now we can plug in to equation 3.3.2 to get $|\kappa| = 1$, or $\kappa = \frac{1}{\bar{\kappa}}$.

We have now arranged for the unit circle to be transformed to itself by $\mathbf{U}$. Since $\mathbf{U}(0) = \frac{\beta}{\delta} = \frac{\beta}{\kappa\bar{\alpha}}$, we need to add the inequality $|\beta| < |\alpha|$. Also, since multiplying all of the coefficients of $\mathbf{U}$ by a real number r doesn't change it, we can adjust α and β so that $|\alpha|^2 - |\beta|^2 = 1$ Putting this all together, we get the answer to our question.

Theorem 3.3.3 *A Möbius transformation* $\mathbf{U}$ *carries the unit disc to itself if and only if there are complex numbers* α, β, *and* κ *with* $|\alpha|^2 - |\beta|^2 = 1$ *and* $|\kappa| = 1$ *for which*

$$\mathbf{U}(z) = \kappa\left(\frac{\alpha z + \beta}{\bar{\beta}z + \bar{\alpha}}\right).$$

Problem

By replacing κ *by* ρ^2 *and using the fact that* $\rho\bar{\rho} = 1$, *show that it is always possible to write* $\mathbf{U}(z)$ *by the formula*

$$\mathbf{U}(z) = \left(\frac{\alpha z + \beta}{\bar{\beta}z + \bar{\alpha}}\right)$$

with $|\alpha|^2 - |\beta|^2 = 1$.

A *fixed point* of a transformation T is a point X for which $T(X) = X$. For transformations in $\mathcal{G}$, fixed points come in pairs:

Problem
Show that if $\mathbf{U}(z) = z$, *for* $\mathbf{U}$ *a Möbius transformation in* $\mathcal{G}$, *then* $\mathbf{U}(z') = z'$, *where* z' *is the inverse of* z *with respect to the unit circle* Ω.

From this last problem and an earlier one, we can see that there are three possibilities for the fixed points of $\mathbf{U}$:

Proposition 3.3.4
If $\mathbf{U}$ *is a Möbius transformation in* $\mathcal{G}$, *then exactly one of the following holds:*

1. $\mathbf{U}$ *has one fixed point inside the unit disc and none on the circle,*
2. $\mathbf{U}$ *has exactly two fixed points on the unit circle,*
3. $\mathbf{U}$ *has exactly one fixed point on the unit circle.*

The first case described in the proposition corresponds to a rotation in $\mathcal{H}$ about some point. The second case arises from a translation in $\mathcal{H}$. The fixed points are the endpoints of the line along which the translation occurs. The third case is a new one. Suppose ω is the fixed point of $\mathbf{U}$. If ℓ is any h-line through ω, then $\mathbf{U}$ takes ℓ to another line through ω. This is just what an ordinary rotation about a point X does to lines through X, so $\mathbf{U}$ is sometimes described as a "rotation" about the ideal point ω. It is called a *parallel displacement about* ω. If ℓ_1 and ℓ_2 are two lines passing through an ideal point ω, and if $\mathbf{F}_1$ and $\mathbf{F}_2$ are the reflections across these lines, then $\mathbf{U} = \mathbf{F}_2\mathbf{F}_1$ is a parallel displacement about ω.

The rotations, translations, and parallel displacements are *direct isometries* of $\mathcal{H}$. If $\mathbf{U}$ is a direct isometry that takes A to A', B to B', and C to C', where A, B, and C are the vertices of a triangle, then not only are ΔABC and $\Delta A'B'C'$ congruent, but the congruence preserves the *orientation* of the triangle. If $\mathbf{V}$ is a reflection or a glide reflection, then the corresponding triangle $\Delta A''B''C''$ has the opposite orientation.

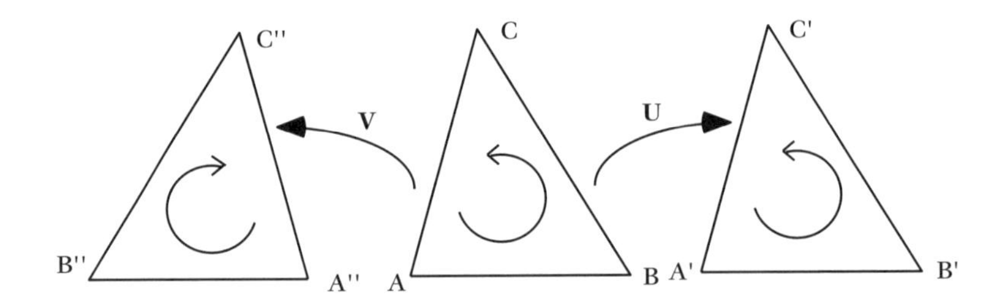

What exactly is orientation? Intuitively, if we slide $\triangle ABC$ around, we can superimpose it over $\triangle A'B'C'$, but we would have to lift it out of the plane and flip it over to put it over $\triangle A''B''C''$. I say "intuitively" because how do we know that someone can't cleverly slide the triangle around to make it fit over $\triangle A''B''C''$? After all, the two triangles are congruent!

We observe that the vertices A', B', and C' are written in order *counter-clockwise* around the triangle, while A'', B'', and C'' appear *clockwise*. But then, what do these words mean? If we are facing a clock that is in a vertical plane, then the hands of the clock move to the right when they are pointing upward and to the left when pointing downward. That seems clear enough, but what do the words "left" and "right" mean? Suppose we had to communicate the notion of clockwise and counterclockwise to someone on another planet, using only audio communication. How could we know that our sense of clockwise would agree with that of people on another planet? Martin Gardner refers to this intriguing question as the "Ozma Problem." ([13], Chapter 18) He quotes William James on this puzzle:

If we take a cube and label one side *top*, another *bottom*, a third *front*, and a fourth *back*, there remains no form of words by which we can describe to another person which of the remaining sides is *right* and which *left*. We can only point and say *here* is right and *there* is left, just as we should say *this* is red and *that* blue.

We know that at least we have been able to adopt a convention that those of us living on this planet will universally understand. How do we know that such an idea makes sense mathematically in a non-Euclidean two-dimensional universe? That is, how can we define orientation in such a way that it consistently holds throughout the plane? A picturesque science fiction version of this question is, Suppose I have a pair of gloves; I leave the left glove at home and go for a long trip carrying the right glove. How can I be sure that when I return home the glove I am carrying will be the opposite of the glove I left behind?

A mathematically rigorous answer is not easy to arrive at. Here is a partial answer. If ℓ is a line in $\mathcal{H}$, then it divides $\mathcal{H}$ into two *half-planes*, defined as follows: Two points X and Y not on ℓ are in the same half-plane if the line segment XY does not meet ℓ. Similarly, a point A on a line ℓ divides ℓ into two *rays*.

Problem

How do we know that there are exactly two half-planes determined by the line ℓ?

Now suppose we fix some half-plane M and fix two points A_1 and A_2 on the line ℓ. If a triangle has its vertices numbered V_1, V_2, and V_3, we will say it is *oriented* by this numbering. It is *positively oriented* if there is an isometry in $\mathcal{G}$ that takes V_1 to A_1, V_2 to a point on the ray from A_1 which contains A_2, and V_3 to a point in the half-plane M. Otherwise, we will say the triangle is *negatively oriented*.

Problem

Show that if $\mathbf{U}(V_i) = (W_i)$ and $\mathbf{U}$ is in $\mathcal{G}$, then the triangles $\triangle V_1V_2V_3$ and $\triangle W_1W_2W_3$ (with the orientations defined by this numbering) have the same orientation. If $\mathbf{U}$ is not in $\mathcal{G}$, then the two triangles have opposite orientations.

Problem

Investigate the idea of orientation in the Euclidean plane and in Euclidean three-dimensional space.

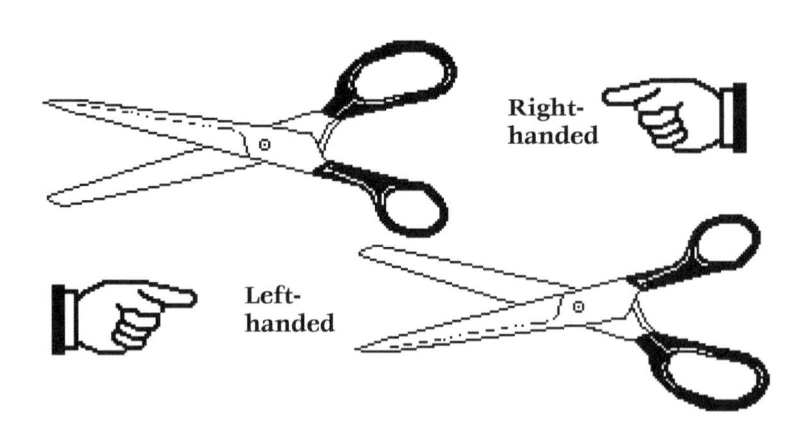

For a fascinating investigation of the problem of orientation, sometimes called *handedness*, see Martin Gardner's book [13]. In Section 4.4, we will encounter a different geometry, in which orientation cannot be defined. In that world, it pays to keep pairs of gloves pinned together during long trips.

CHAPTER 4 Geometry of the Sphere

4.1 Spherical Geometry as Non-Euclidean Geometry

Euclid's axiom, as formulated by Hilbert (see Section 1.1), states that given a point A not lying on a line a, there is at most one line passing through A that does not intersect a. Hyperbolic geometry replaced that axiom with the assumption that more than one line through A does not intersect a. These are the only two possibilities consistent with the remaining axioms. From Hilbert's axioms we can always construct one line through A not meeting a.

Now we want to consider a third geometry, in which through a point A not on a line a there is *no* line that does not meet a. This is not possible from Hilbert's other axioms, for reasons that will be easier to see a little later. If we look at this assumption as a replacement for Euclid's axioms, it still appears impossible. For one thing, it implies the existence of triangles whose angle sums are greater than 180°, which Theorem 1.3.2 says is impossible. Let's examine one of these impossible triangles.

Let A and B be two points lying on a line c. Construct the line a through B perpendicular to c. Now construct the line b through A perpendicular to c. If we assume that this line meets a at some point C, then the resulting triangle $\triangle ABC$ would have two right angles. (It is necessary to draw curved lines in the figure below to represent straight lines. Try to imagine these as straight lines.)

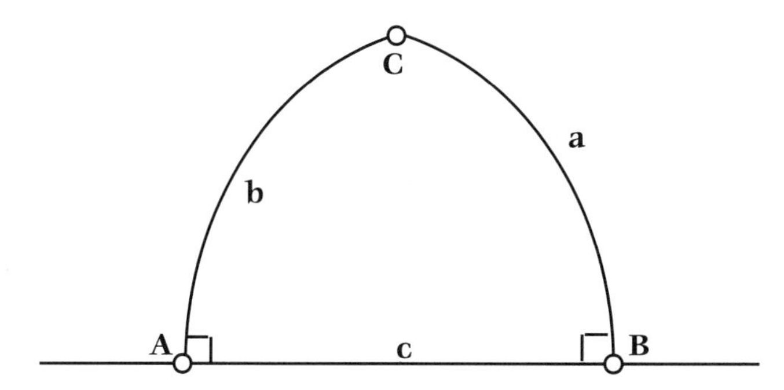

But wait; it gets worse. Choose a point A' on the line c such that $|BA'| = |AB|$. Then the triangle $A'BC$ must be congruent to ABC; this means that the segment CA' is also perpendicular to c. All three of these perpendiculars must have the same length, since we can easily prove that ABC is isosceles. We can keep extending the line c and constructing more perpendiculars from C. (But it gets harder to draw a convincing picture of this!)

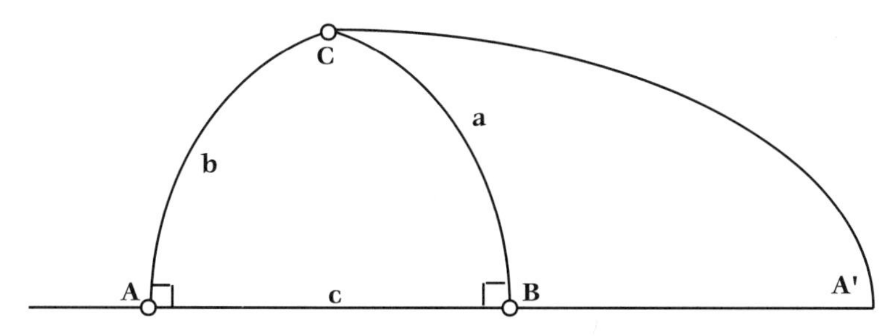

We also can construct $\triangle ABC'$ congruent to $\triangle ABC$ on the other side of the line c. Now, since the angles at A and B are right angles, these two triangles fit together in such a way that C, A, and C' all lie on a line, and also the points C, B, and C' all lie on a line. Now we seem to have two straight lines passing through the points C and C', forming a "biangle", a polygon with two sides!

It is pretty easy to see from all of these pictures why this geometry was not considered a possibility even after the discovery of hyperbolic geometry. Then, in 1854, Georg Friedrich Bernhard Riemann presented a lecture entitled "Über die Hypothesen welche der Geometrie zu Grunde liegen" (On the Hypotheses Which Lie at the Foundation of Geometry). At the time, Riemann was just a lowly lecturer, or *Privatdozent*, at the great German University at Göttingen, and it was customary for such a lecturer to give an inaugural presentation to the faculty. This lecture was

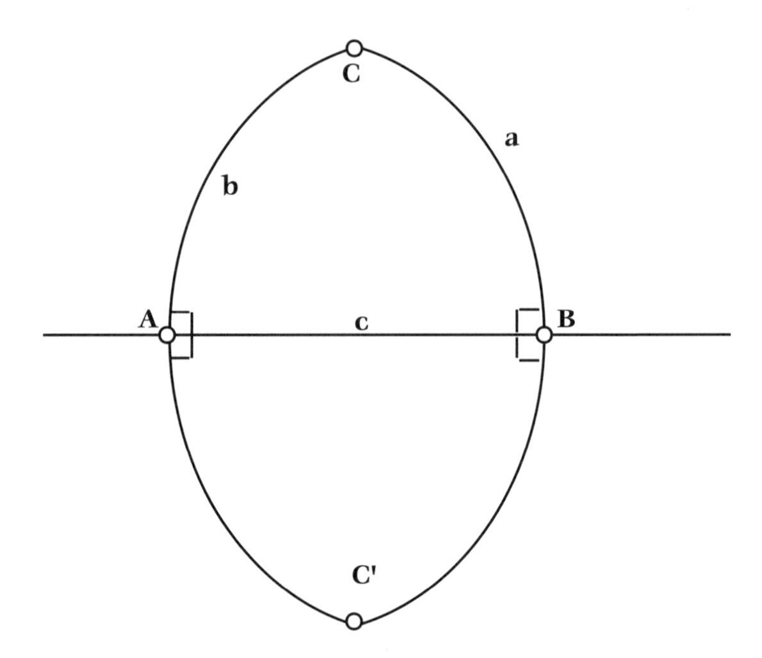

arguably the most significant inaugural lecture in history: It transformed our view of geometry forever. (An English translation of the lecture can be found in [32], vol. 2. It is well worth reading.)

We will look at Riemann's ideas about geometry and space more carefully in Chapter 6. For now, let us zero in on one small piece of his theory. I quote the relevant passage from [32]. The meaning of some of the words may not be clear, particularly the words "manifold" and "curvature." These will be discussed in Chapter 6.

> When constructions in Space are extended into the immeasurably large, unboundedness is to be distinguished from infinitude; one belongs to relations of extension, the other to metric relations. That space is an unbounded triply extended manifold is an assumption which is employed for every apprehension of the external world, by which at every moment the domain of actual perception is supplemented, and by which the possible locations of a sought for object are constructed; and in these applications it is continually confirmed. The unboundedness of space consequently has a greater empirical certainty than any experience of the external world. But its infinitude does not in any way follow from this; quite to the contrary, Space would necessarily be finite if one assumed independence of bodies from position, and thus ascribed to it a constant curvature, as long as this curvature had ever so small a positive value. If one prolonged the initial

> directions lying in a surface into shortest lines, one would obtain an unbounded surface with constant positive curvature, and thus a surface which in a flat triply extended manifold would take the form of a sphere, and consequently be finite.

What Riemann was saying is that the idea that space is *unbounded* is different from the idea that it is *infinite*. A straight line in Eucld's geometry is infinite. A circle is finite, but it has no beginning or end. Why not imagine a geometry in which a straight–line actually closes up on itself like a circle? One could travel in a straight line direction forever and not come to the end of it. But maybe if we did that we would come back to where we started over and over again.

The simple model Riemann proposed for this is a sphere. In this model of geometry, called *spherical geometry*, a **point** is a point on S, the surface of the sphere of radius 1; and a **line**, which we will call an *s-line*, is a great circle. (Any plane intersects the sphere in a circle; a plane through the origin intersects the sphere in a great circle.) Angles are measured in the usual (Euclidean) way. In order for this model to fit into the framework of geometry that we have been using, we have to modify Euclid's first postulate to mean that there is *at least* one straight line passing through any two points. If the two points are not *antipodal points* on the sphere, then there will be exactly one s-line. But for example, every meridian on the globe corresponds to (half of) an s-line passing through the north and south poles of the sphere. The second postulate must be understood that a line has no limit, not that it is infinite. The fifth postulate, of course, is to be replaced by the postulate that any two straight lines intersect (in fact, in a pair of antipodal points). If A and B are not antipodal points, then there is a unique *s-line segment* joining them that is a piece of a great circle not containing any pair of antipodal points. There is no way to specify a line segment uniquely between a pair of antipodal points. Consequently, arguments about triangles become very complicated unless they are *small*.

What is a triangle on the sphere? Suppose A, B, and C are points in the sphere S that do not lie on the same s-line. In particular, no two of them can be antipodal. If c is the s-line determined by A and B (which is unique, because A and B are not antipodes), then c divides S into two hemispheres, one of which contains C. We get a unique spherical triangle ABC, which is contained in this hemisphere. It is not too difficult to prove that an s-line that passes through a point of the segment AB must also pass through a point of either AC or BC. So Pasch's axiom, Hilbert's Axiom II,4, holds in S.

Problem

Examine the other Hilbert axioms of geometry (Section 1.1). Which of them hold in spherical geometry? You will need to define "between" somehow in order to do this problem.

We have not yet defined congruence in our model for geometry. The answer comes from three-dimensional Euclidean geometry. A rotation of three-dimensional Euclidean space around some line through the origin takes points on the unit sphere to points on the unit sphere. It preserves angles between curves, since it is an isometry of Euclidean geometry. Since it also takes planes through the origin to planes through the origin, it takes great circles to great circles and preserves angles between them. In addition to the rotations, there are reflections. If P is any plane through the origin, then reflection through this plane will take the unit sphere to itself, fixing points on the great circle that is the intersection of P with S. This is defined to be a reflection in S.

Just as we used complex analysis to write down formulas for isometries, we will find formulas for the isometries of S. This turns out to be a formidable task, however, so let's put this off until Chapter 5.

But what about our theorem that says there is no such geometry? If we assume that a line can be unbounded without being infinite, then we can find a flaw in Legendre's argument that the sum of the angles in a triangle cannot be greater than $180°$. The argument assumed that if we travel on a straight line between two points, then we must be following a shortest path. In fact, on a sphere, a great circle arc is the shortest path that stays on the sphere and joins the two endpoints, *provided* that it does not contain antipodal points. In the proof of Theorem 1.3.2, the possibility that the path joining A_1 to A_2 to A_3, etc., might not be the shortest path was not considered.

On the sphere it is actually quite easy to construct a triangle with two right angles. If A and B are any two points on the equator and C is the north pole, then ΔABC has two right angles. All of the pictures from the beginning of this section make perfectly good sense in S. In fact, by choosing AB to be a quarter circle, we may construct an equilateral triangle with three right angles. Eight of them fit together perfectly to fill out the sphere. This is our first example of a regular tessellation of the sphere.

Just as we constructed regular polygons in hyperbolic space by picking points on rays from some central point, we can construct regular polygons on the sphere. To construct a regular hexagon, for example, draw six meridians from the north pole N at $60°$ angles. A plane perpendicular to the line through the north and south poles will cut the sphere in a small circle that crosses each meridian at a point. Now connect consecutive points with s-lines. Since the rotation around the north–south axis is an isometry of S, the polygon is regular. An important consequence of the construction is worth noting:

Proposition 4.1.1
The vertices of a regular spherical polygon lie in a plane in Euclidean space. Therefore, the vertices of a regular spherical polygon are the vertices of a regular Euclidean polygon that is inscribed in the sphere.

To determine the interior angles of the regular polygon, we use an analysis similar to that used in the hyperbolic case. The hexagon is made up of six isosceles triangles, each having an angle of 60° at its apex. Only this time, instead of the sum of the angles being less than 180°, it turns out to be larger. Consequently, we define the *excess* of a spherical triangle to be 180° minus the sum of the angles in the triangle.

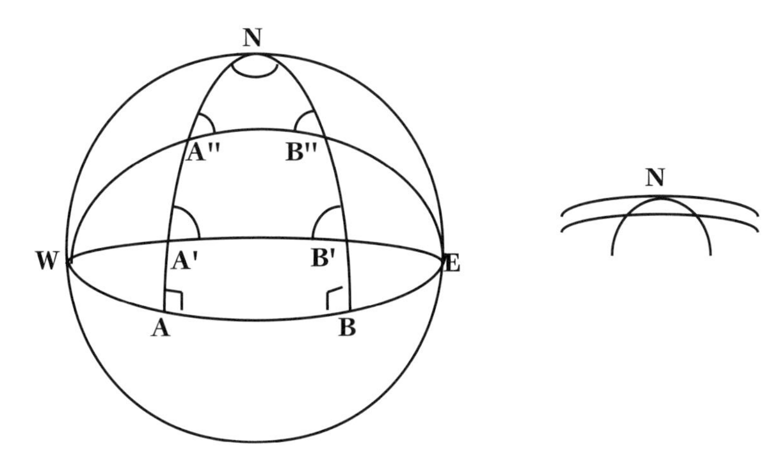

Each isosceles triangle is made from two meridians from the north pole N together with a great circle that passes through a pair of antipodal points E and W on the equator. In the special case where the third side is part of the equator, the resulting triangle ΔNAB has base angles of exactly 90°. If the triangle $\Delta NA'B'$ has its vertices in the northern hemisphere, then the base angles will be smaller than 90°. As we move the base to points A'' and B'' closer to N, the angles continue to decrease. When they are very close to N, the picture looks like a tiny triangle with straight sides; in fact, the base angles decrease to the Euclidean values of 60°. To prove this, the key is to notice that the spherical triangle can be more and more closely approximated by the (planar) Euclidean triangle with the same vertices as it shrinks in size. There is a more general fact that is true: By exact analogy with the hyperbolic case, it is easy to prove that the excess is additive. Remember that this means that if a triangle is cut up into smaller triangles, then the excess of the triangle is the sum of the excesses of the smaller triangles. Using this fact, it is not too difficult to prove:

Lemma 4.1.2
The area of a spherical triangle is proportional to its excess.

Since the triangles we are examining sit inside a spherical lune that has exactly one-sixth the area of a sphere, we can use this example to determine the constant of proportionality. For the right triangle NAB has area exactly $\frac{\pi}{3}$ ($\frac{1}{12}$ the total area of the unit sphere) and has excess $90 + 90 + 60 - 180 = 60$. So the constant of proportionality must be exactly $\frac{\pi}{180}$.

Incidentally, this is an illustration of the advantages of using radian measure for angles. If we were measuring angles in radians, the triangle would have angles $\frac{\pi}{2}$, $\frac{\pi}{2}$, and $\frac{\pi}{3}$, and its area would be exactly equal to its spherical excess.

Now, since the isosceles triangle sits inside the lune whose area is $\frac{2\pi}{3}$, its spherical excess can not be any larger than 120. Therefore the base angles must lie strictly between 60° and 120°. The interior angles of a regular hexagon are therefore between 120° and 240°. When the interior angles get bigger than 180°, then of course, the hexagons start to be a bit weird. In fact, the hexagon with vertex angles exactly 180° is really just a hemisphere. Let's rule out angles this big in our regular polygons. Then the following theorem describes the situation on the sphere completely:

Theorem 4.1.3 *There is a regular polygon with n sides in the sphere whose interior angles are $\theta°$ for $\frac{180(n-2)}{n} < \theta < 180$.*

Now it is routine to find the regular tilings of the sphere, using our familiar methods. Hexagons don't work, since even three of them will no longer fit around one vertex. It is possible to fit three pentagons around one vertex, but not four. It is possible to fit three regular quadrilaterals around one vertex, but again not four. With triangles it is possible to fit three, four, or five around a vertex. In all, this gives us five possibe tilings of the sphere.

How can we see that in each case, it is possible to complete the construction? By Proposition 4.1.1, if we have such a tiling, we can inscribe regular polygons whose vertices correspond to the vertices on the sphere. These will fit together to give an object made up of polygonal sides, a *polyhedron*. Now in fact, there are exactly five polyhedra made up of identical regular polygons: These are the five *Platonic solids*.

The five solids—the tetrahedron, cube, octahedron, dodecahedron, and icosahedron—are constructed in the final propositions of Book XIII of Euclid's *Elements* (see Heath ([19]). They appear to have been first studied by Theaetetus; Plato viewed them as being central to the scientific study of the universe.

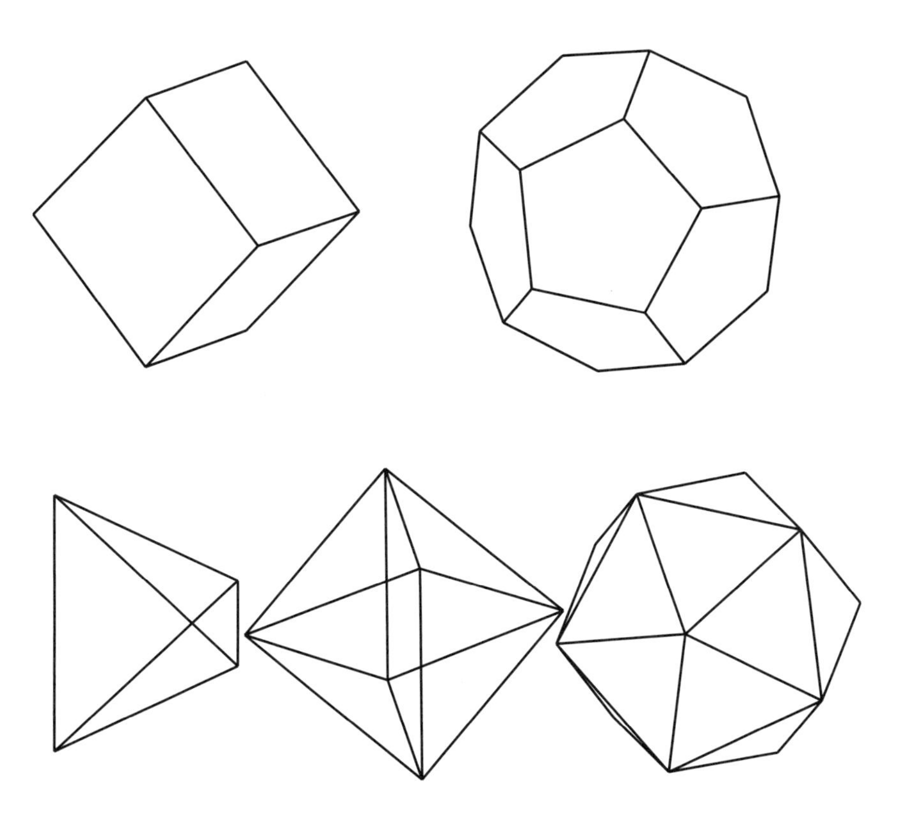

It turns out to be very instructive to note certain statistics concerning the platonic solids. Let's denote by V the number of vertices (corners) in one of these polyhedra, by E the number of edges, and by F the number of polygonal faces. The results are summarized in the following chart:

	V	**E**	**F**
Tetrahedron	4	6	4
Cube	8	12	6
Octahedron	6	12	8
Dodecahedron	20	30	12
Icosahedron	12	30	20

The similarity in the statistics for the octahedron and the cube, and for the dodecahedron and icosahedron, is easily explained. As in the Euclidean and hyperbolic cases, we can form the dual tiling to each of the regular tilings of the sphere. The dual tiling will have one vertex for each face of the original tiling, one face for each vertex, and one edge for each edge. The cubic and octahedral tilings are duals, as are the dodecahedral and icosahedral tilings. The tetrahedron is its own dual, so $V = F$.

Another statistic is more subtle: In each case,

$$V + F = E + 2.$$

This is the content of *Euler's theorem*, and it is a powerful tool in the study of tilings of the sphere, because it is true for *any* tiling. In the next section, we will examine this theorem and its proof; its consequences will appear in Section 4.3 as well as in Chapter 5.

4.2 Graphs and Euler's Theorem

The formula of Euler was stated and proved by Leonhard Euler in 1752. In 1860 a partial manuscript written by René Descartes, containing a theorem about polyhedra that can be used to prove Euler's formula, was discovered among the papers of Leibniz [3], so many scholars concluded that Descartes knew the formula as early as 1635, and the formula is known by some as the Euler–Descartes formula. According to Eves ([11], p. 74), it may even have been known to Archimedes. However, most recent scholarship disputes these claims and returns the recognition to Euler. (See the excellent article by Joseph Malkevitch in [31], pp. 80–92, for a historical review of the formula and of polyhedra in general.) It is one of the fundamental results in the area of mathematics known as *algebraic topology*, where it has been generalized in a large number of ways.

The proof of the formula is relatively easy, although this is a bit deceptive. There are some technical difficulties lurking in the undergrowth, which we will try to avoid stepping on. The most significant of these is something called the *Jordan curve theorem*. This innocent-sounding theorem says that a simple closed curve C in the plane (or on the sphere S) divides it into two regions. The word "simple" refers to the fact that C has no self-intersections. The word "divides" refers to the fact that a point in one region can be connected to any other point in the same region by a curve that does not touch C, but any curve joining points from different regions must cross C.

The difficulty in proving this theorem comes from the fact that a simple closed curve can be very complicated; for example, it can be a fractal curve such as the boundary of the Koch snowflake or the pentaflake from Chapter 2. If the curve is not so awful, for example if it is made up of finitely many straight line segments in the plane or circle arcs on the sphere, then the technical difficulties are not too great. For a relatively elementary proof, see [34], pp. 26–35.

We will take the Jordan curve theorem for granted here. Next, we need some terminology. A *graph* on the sphere or the plane consists of a finite collection of points called *vertices* and a finite collection of curves called *edges*; each endpoint of an edge should be a vertex. It is permissible to have both ends of an edge at the same vertex; this is called a *loop*. It is permitted to have two edges connecting the same pair of vertices. It is also permitted for two edges to cross, but we will require that they cross at no more than one point. Edges that each have a vertex x as an endpoint are called *incident*; these should not cross. No edge should pass through a vertex.

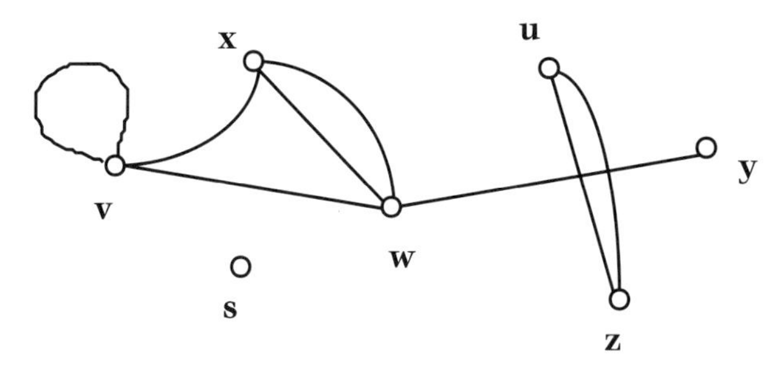

Two vertices a and b of a graph are said to be in the same *component* if there is a sequence of vertices $(a_1, a_2, a_3, \ldots, a_n)$, where $a_1 = a$, $a_n = b$, and each pair of consecutive vertices in the list is connected by an edge. In the graph pictured above, there is a sequence (v, x, y) connecting v to y, so they are in the same component. There is no such sequence connecting v and z. The components of the graph are the vertices $\{v, w, x, y\}$, the vertices $\{u, z\}$, and the single vertex $\{s\}$. (We have to specify that every vertex is in the same component as itself to get that last case.)

We will mainly be interested in graphs that have the added property that no two edges cross; such a graph is said to be *embedded*. Suppose G is such a graph. Points in the plane (or the sphere) that are not on any of the edges of the graph are divided up into regions called *faces*. Two points are in the same face if they can be joined by a path that does not touch the graph. Let $V = V(G)$ be the number of vertices of the embedded graph G, E the number of edges, F the number of faces in the plane or sphere determined by the graph, C the number of components of G.

Theorem 4.2.1 (Euler's Theorem) *For any embedded graph G in the plane or sphere,*

$$V - E + F = 1 + C.$$

Proof

The idea of the proof is an induction argument. If there are no edges in the graph, then $E = 0$, $F = 1$, and $V = C$. Now assume that the theorem is true for any graph with $E = n$. Suppose G is a graph with $n + 1$ edges. Pick an edge e of the graph joining two vertices a and b, and delete it. Call the new graph G'; this graph has n edges. Then $V(G') = V(G)$, $E(G') = E(G) - 1$, and $V(G') - E(G') + F(G') = 1 + C(G')$ hold by our assumption that the theorem is true for graphs with n edges. There are several possibilities:

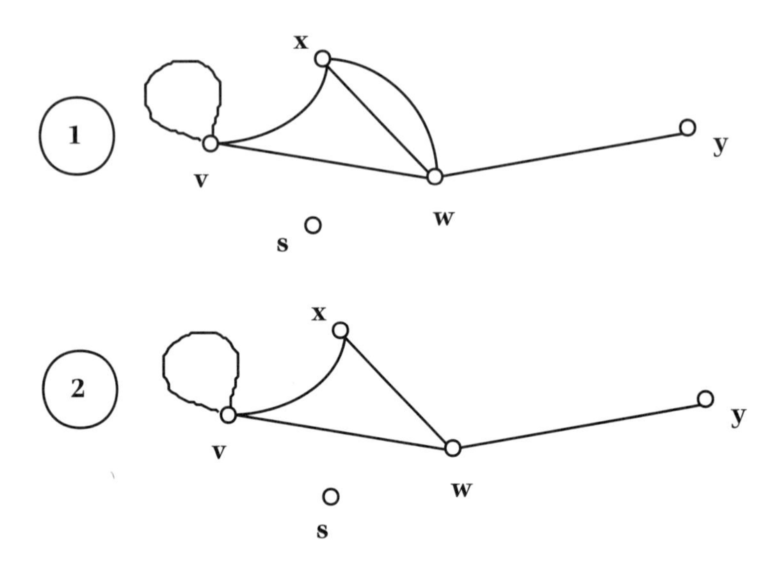

If the vertices were connected by more than one edge, then $C(G') = C(G)$ and $F(G') = F(G) - 1$. This is illustrated in the transition from graph (1) to graph (2). More generally, if in graph G' the vertices a and b are in the same component, then there is a path of edges from a to b that when combined with the deleted edge e between a and b forms a simple closed curve. By the Jordan curve theorem, points on opposite sides of e cannot be connected by a path that does not touch the graph G. But after deleting e we can easily connect two such points. Therefore, the number of faces must have decreased, and $F(G') = F(G) - 1$. Since there is still a path joining a and b, $C(G') = C(G)$. The transition from (2) to (3) illustrates this case.

The final possibility is the most technically subtle. If a and b are in different components of G', then $C(G') = C(G) + 1$. In this case, which is illustrated in the transition between (3) and (4), the number of faces does not change: $F(G') = F(G)$. But how do we prove this? The fact that points on opposite sides of the deleted edge can be connected by a path that does not touch G can be seen in the picture, but proving that it exists is hard. Fortunately, if we are careful, we can avoid this problem.

If we are careful, we can always choose an edge for which one of the vertices has no other edges connected to it. Begin with the edge we first chose and follow a path of edges going from one vertex to another, always leaving by a different edge. Eventually, one of two things will happen. If we come back to a vertex we have already visited, then we have found

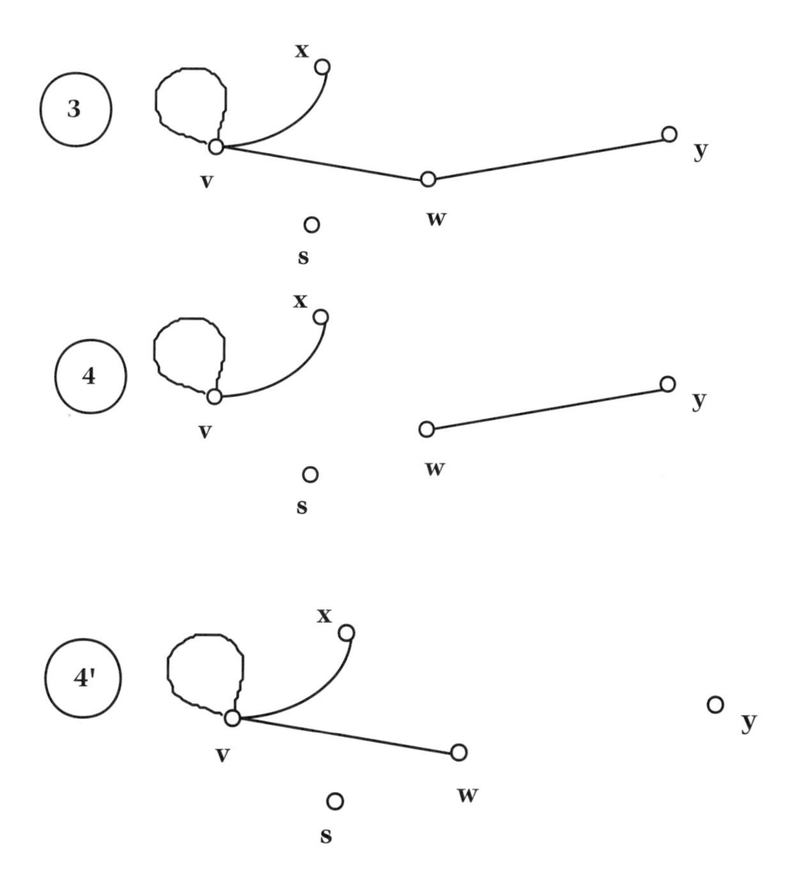

a closed curve of edges. In that case, deleting one of these edges does not disconnect the vertices. This is the case we have already dealt with. The second possibility is that we can no longer move. In that case, the last edge must have a vertex with no other edges on it. If we delete that edge, then we have $C(G') = C(G)+1$ and $F(G') = F(G)$. This last equation comes from the fact that points on opposite sides of the deleted edge e can be connected by a path that walks around the free end of the edge. (Technically, this is the *Jordan arc theorem*, and its proof is as subtle as the Jordan curve theorem. But if the arc is not an ugly curve, this is not hard to prove.)

In each of the cases above, one number on each side of the equation changes by 1 in such a way as to keep the balance. So if the equation holds for G', it must also hold for G. ■

A useful exercise is to continue deleting edges in the example above until there are none left, while computing V, E, F, and C at each stage.

Problem

If the graph is not embedded, and X denotes the number of points where two edges cross, what is the relationship of V, E, F, C, and X? What happens if three edges go through the same point? (For simplicity, assume $C = 1$; that is, assume that any two vertices are connected by a legitimate path of edges. Test your result on the pentagram.

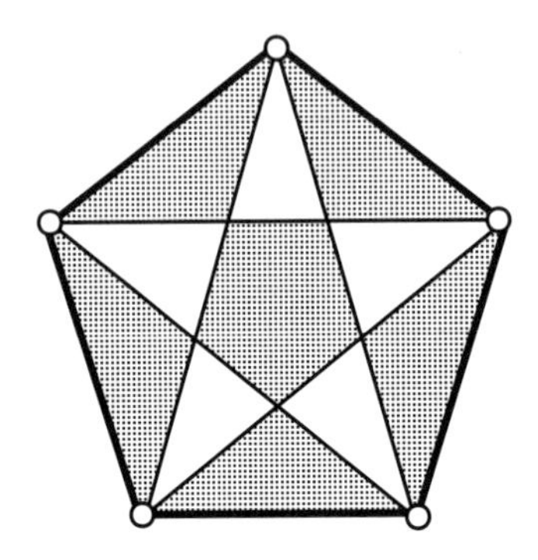

As an application of Euler's theorem, consider the following problem: Suppose a map of Europe shows five countries, each of which has a border consisting of a simple closed curve. Is it possible that each pair of countries shares a common border? We assume that a common border consists of more than one point, otherwise we could easily have five countries shaped like wedges of pie all touching at one point.

To answer this, we use the technique described in Chapter 2 of forming the *dual* of the map. Suppose we choose a point in the middle of each country and draw curves between these points that connect pairs of countries across their common borders. The result is a graph with five vertices, one for each country, and ten edges, one for each border crossing. We can draw these curves so that no edges cross. By Euler's formula, $F = 2 - V + E = 7$, so this graph must divide the sphere into seven regions. But each region must have at least three sides, for otherwise we would have two edges connecting the same two vertices. Since there are seven regions, that makes a total of at least 21 sides. Each of the ten edges accounts for two sides, so we can only account for 20 sides. So this map doesn't exist! Stated another way, if we put five points in the plane or sphere and connect each pair of points with an edge, then at least one pair of edges must intersect.

Problem (The Utilities Puzzle)

Three utilities provide water, phone, and electricity, respectively, to houses in a community. Each utility wants to dig a trench and run a conduit from its plant to each of three houses. Show that two of these trenches are going to have to cross. (Note: The companies are not allowed to cheat by connecting one house directly to another!)

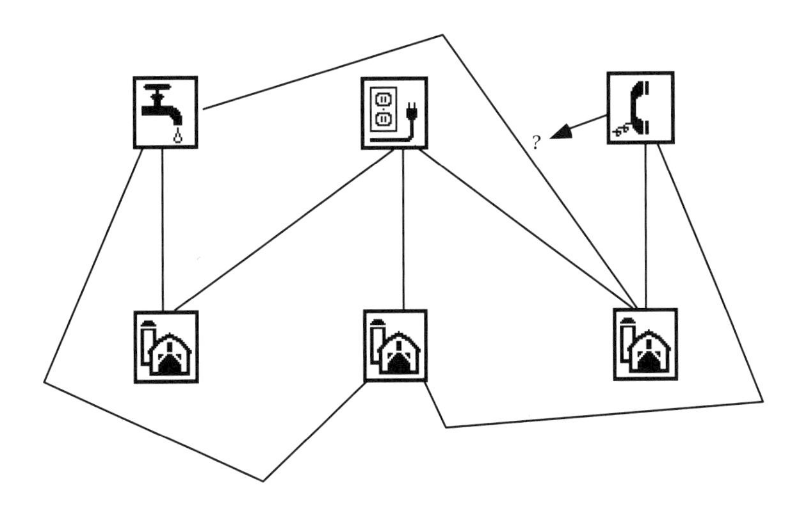

Coloring. When designing maps showing different countries, states, or other political subdivisions, it is desirable to use different colors for different regions in order to make it easy to distinguish them. The same color may be used for more than one country, but different colors are used for adjacent countries to delineate borders. We have shown that it is not necessary to use five different colors to distinguish five different countries.The *four color theorem* states that in fact, it is possible to color any map on the sphere or plane using only four different colors in such a way that no two adjacent countries have the same color. (This is only true if countries are assumed to have borders that are simple closed curves, as above. A country that is made up of separate pieces may have to have different parts colored with different colors.) We also color oceans, etc.

That four colors are necessary can be seen in many examples; for instance, if we color the ocean blue and try to color the states of the U.S., we need different colors for Florida, Georgia, and Alabama, none of which

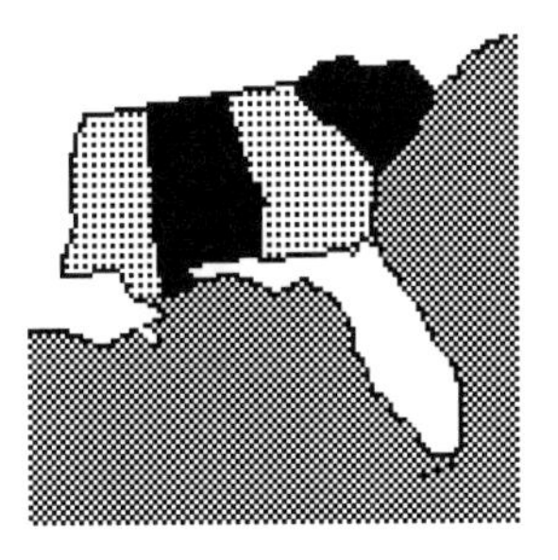

can be colored blue. We do not need a fifth color for Mississippi or South Carolina.

The four color problem has a long and interesting history. It dates back at least to 1840, when it was mentioned by the mathematician A.F. Möbius, and it was an unsolved problem until 1976, when Ken Appel and Wolfgang Haken of the University of Illinois announced their proof. The proof itself is an enormous one, involving a computer analysis of hundreds of special cases. Needless to say, the complete proof is not an easy one to read. There is a very readable discussion of the four color theorem, including a brief explanation of the main ideas of the Haken-Appel proof, in [3]. This reference has quite a lot of interesting material on polyhedra, Euler's theorem, and graphs. It also has the solution to the following interesting (and difficult) puzzle:

Problem (The Colony Problem)
Suppose we have a map in which each country has at most one colony. *Each country has a border that is a simple closed curve, and its colony also has such a border. We want to color the map so that no countries with a common border have the same color, but any colony has the same color as the country to which it belongs. How many colors suffice to color any such map?*

Although the four color theorem is difficult to prove, we can easily prove that *five* colors are sufficient to color any map. The techniques used here will be very helpful in the next section. Suppose there is a map that cannot be colored with five colors. We can choose such a map with the smallest number of countries; call such a map "critical". Corresponding to the map is a graph G with a vertex for each country and an edge for each common border. The number of sides of all faces determined by the graph is at least $3F$, since each face must have at least three sides. (We are counting the "outside" region, which goes off to infinity, if we are on the plane. On the sphere it is just another region.) Since each edge is a side for two regions, this gives the inequality

$$3F \leq 2E.$$

Euler's theorem tells us that $V - E + F = 1 + c \geq 2$, so combining these inequalities,

$$2 \leq V - E + \frac{2}{3}E = V - \frac{1}{3}E,$$

or

$$E \leq 3V - 6. \tag{4.2.1}$$

Since every edge has two ends, the average number of edges per vertex is $2E/V$. By equation 4.2.1

$$\frac{2E}{V} \leq 6 - \frac{12}{V} < 6.$$

This tells us that *there is a country that has no more than five neighbors.* Suppose some country has no more than four neighbors. Let one of its neighbors invade and annex it. Now we redraw the map; it has one less country, so it is *subcritical.* It has too few countries to be a problem, so by assumption we can color it with five colors. Now redraw the conquered country. Since it only touches four other countries, we can recolor it using the fifth color.

Suppose instead that a country touches five other countries. By our earlier theorem, there must be two among these five countries that do not share a common border. This time, imagine that the country in the middle conquers both of these countries. The new map is subcritical, so we can color it. Now divide that country back into three again. The one in the middle has five neighbors, but only four colors were used to color them. So we can color the middle one with the fifth color.

The conclusion of the argument is that our so-called "critical" map isn't critical after all. That contradiction proves the theorem.

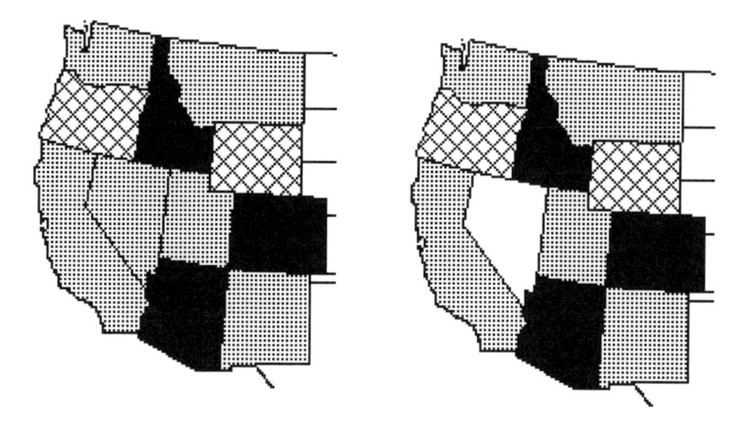

Nevada is surrounded by exactly five other states. To color the map of the U.S., first let Nevada annex California and Utah. Color the map. Now notice that there are in this example only three colors used to color Nevada's neighbors. Pick one of the available colors (in this example, Nevada gets the same color as the Pacific Ocean.)

Problem

Modify the argument above to show that any map that has no more than 11 regions can be colored using only four colors.

Problem

Show that the dodecahedron can be colored with four colors. If a map has 12 regions and every region touches five other regions, show that it must look like the dodecahedron map. Conclude that any map with no more than 12 countries can be colored with four colors.

4.3 Tiling the Sphere: Regular and Semiregular Polyhedra

In searching for all regular and semiregular tilings of the plane, we relied on the fact that only certain regular polygons fit together around a point so that the sum of the angles around a point was exactly 360°. If we did not insist on regular polygons, this technique would not have been available to us. In fact, we saw that it is possible to tile the plane with copies of a (nonregular) pentagon, while regular pentagons do not fit together in the plane.

Euler's theorem tells us that the situation is somewhat different on the sphere. Suppose, for example, we wish to tile the sphere with pentagons, without concerning ourselves about whether they are regular polygons. We ask only that exactly k pentagons come together at each vertex, where k is a whole number. If F is the number of polygons, then the number of edges must be exactly $\frac{5}{2}F$, while the number of vertices must be exactly $\frac{5}{k}F$. Euler's equation implies that $V = 2+\frac{3}{2}F$, so eliminating V and solving for F in terms of k gives us

$$F = \frac{4k}{10-3k}.$$

Since k can't be less than 3, the only possible value for k is 3, and $F = 12$. With a little patience we can check that the only way to fit 12 pentagons together is in the pattern given by the dodecahedron. One way to describe this pattern is to use stereographic projection to draw this in the plane; the result is a graph, known as a *Schlegel diagram*, consisting of a pentagon divided into 11 smaller pentagons. This is the map of 12 countries (including the outer one) each of which touches five others.

Problem

Can you find a nonregular pentagon, copies of which fit together to fit a nonregular dodecahedron?

Suppose the sphere is divided into polygonal regions in such a way that every region has n sides, and k polygons come together at each vertex. Call this a *regular subdivision* of the sphere.

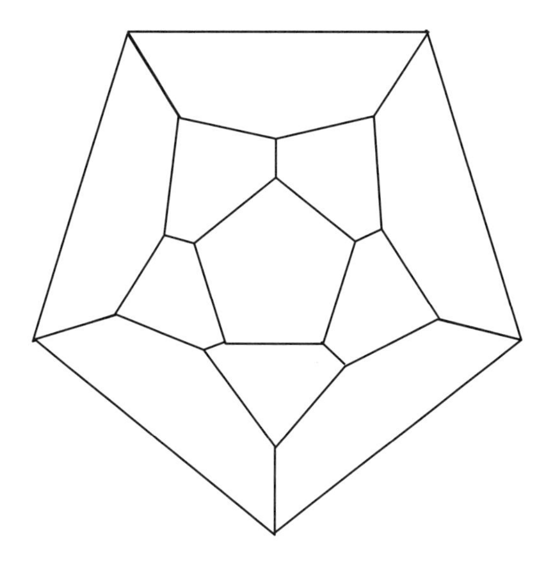

Proposition 4.3.1

There are exactly five pairs of numbers $\{n, k\}$ for which there exist regular subdivisions of the sphere. These correspond to the five regular polyhedra. For any regular subdivision, the corresponding graph in the plane is uniquely determined.

Proof

If a regular subdivision has V vertices, E edges, and F faces, then counting the total number of sides to all the regions in three different ways gives

$$2E = nF = kV.$$

By Euler's theorem, $2kV - 2kE + 2kF = 4k$, so $2nF - knF + 2kF = 4k$. Solving for F and fiddling with the denominator yields

$$F = \frac{4k}{2n - kn + 2k} = \frac{4k}{4 - (n-2)(k-2)}.$$

Since F is positive, we must have $(n - 2)(k - 2) < 4$. There are exactly five solutions to this inequality, corresponding to the five regular polyhedra. For each one, the value of F is uniquely determined, as are V and E. It is not hard to check that each of the five determines uniquely a simple graph in the plane. ■

The same type of argument can be used in narrowing down our search for possible semiregular tilings of the sphere. Given a tiling of the sphere, we have a graph G consisting of the vertices and edges of the tiling. Using stereographic projection, we can think of the graph as being an embedded graph in the plane. Call such a graph *uniform* if there is the same number k of edges at every vertex, and for every m there is the same number of m-sided faces at each vertex.

Suppose that at each vertex there are exactly f_3 three-sided faces, f_4 four-sided faces, and so on. Suppose there are V vertices. The number of edges at each vertex is the same as the number of faces, so the number of edges in the graph is given by

$$2E = (f_3 + f_4 + f_5 + \cdots)V.$$

The total number of triangles is $\frac{f_3 V}{3}$. We have to divide by three because we count each one three times. Likewise, there are $\frac{f_4 V}{4}$ four-sided faces, etc. Adding these up, we obtain

$$F = V\left(\frac{f_3}{3} + \frac{f_4}{4} + \frac{f_5}{5} + \cdots\right).$$

By Euler's theorem,

$$\begin{aligned} 4 = 2V - 2E + 2F &= 2V - V\left[f_3 + f_4 + f_5 + \cdots\right] \\ &\quad + V\left[\frac{2f_3}{3} + \frac{2f_4}{4} + \frac{2f_5}{5} + \cdots\right], \qquad (4.3.1) \\ 4 &= V\left[2 - \frac{1}{3}f_3 - \frac{2}{4}f_4 - \frac{3}{5}f_5 - \cdots\right]. \qquad (4.3.2) \end{aligned}$$

The quantity inside the brackets in equation 4.3.2 must be positive, which means that

$$\frac{1}{3}f_3 + \frac{2}{4}f_4 + \frac{3}{5}f_5 + \frac{4}{6}f_6 + \cdots < 2. \qquad (4.3.3)$$

Lemma 4.3.2

There are at most three different types of region at any vertex of a uniform graph.

Proof

If four of the terms on the right side of equation 4.3.3 were nonzero, then they would have to add up to at least $\frac{1}{3} + \frac{2}{4} + \frac{3}{5} + \frac{4}{6} = 2.1$, contradicting the inequality. ∎

That narrows down our search for uniform graphs a bit. Equation 4.2.1 helps us a lot more; it says that the average number of edges at each vertex is less than 6. That means that if $f = f_3 + f_4 + f_5 + \cdots$, then $f = 3$, 4, or 5. Now, for each of the possible values of f we can study the solutions to equation 4.3.2. For example, suppose $f = 3$ and there are three different polygons at each vertex. If $a < b < c$ are the numbers of sides of the three polygons, then we have

$$4 = V\left[\frac{2}{a} + \frac{2}{b} + \frac{2}{c} - 1\right]. \tag{4.3.4}$$

None of the three numbers a, b, and c can be odd. For suppose, say, b were odd. Walk around the face with b sides; the type of face adjacent to each side would have to alternate. But then when we get back to where we started, the type doesn't match!

Now, we have $a \geq 4, b \geq 6, c \geq 8$. There are two possible solutions to equation 4.3.4. Each of these turns out to correspond to a polyhedron. One of them, which has the tiling pattern $\{4, 6, 8\}$ (to use the notation of Chapter 2), has $V = 48, E = 26, F = 72$; it is known as the *great rhombicuboctahedron*. The other, $\{4, 6, 10\}$, has $V = 120, E = 62, F = 180$; it is called the *great rhombicosidodecahedron*.

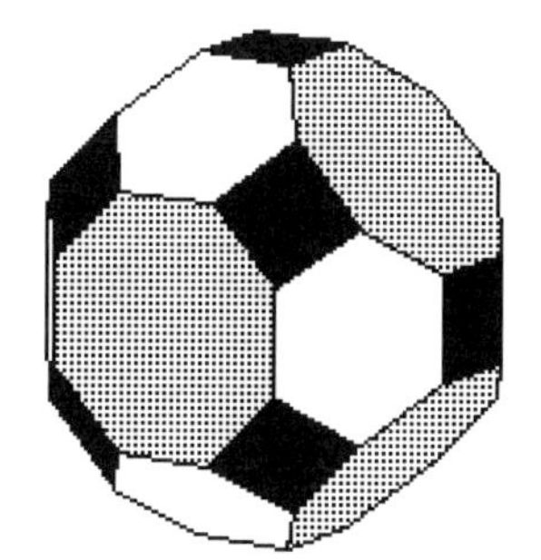

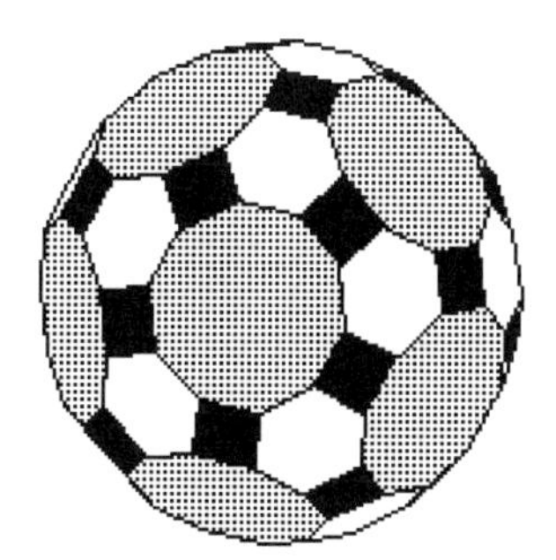

Suppose $f = 3$, and there are two polygons with a sides and one with b sides at each vertex. Then a calculation like the one above gives

$$4 = V\left[\frac{4}{a} + \frac{2}{b} - 1\right].$$

If a were an odd number larger than 3, we would run into the same type of difficulty described above; b can be odd, however. There are actually infinitely many combinations that work. If $a = 4$ and b is any even number bigger than 6, we can find a *prism* consisting of two regular polygons with b sides connected by squares. When $a = 3$ and b is an even number greater than two, there is an *antiprism* with two b-sided polygons joined by equilateral triangles.

Apart from the prisms and antiprisms, there are just five uniform graphs that have three edges at each vertex: the *truncated tetrahedron*, the *truncated cube*, the *truncated octahedron*, the *truncated dodecahedron*, and the *truncated icosahedron*. These arise in a very simple way from the

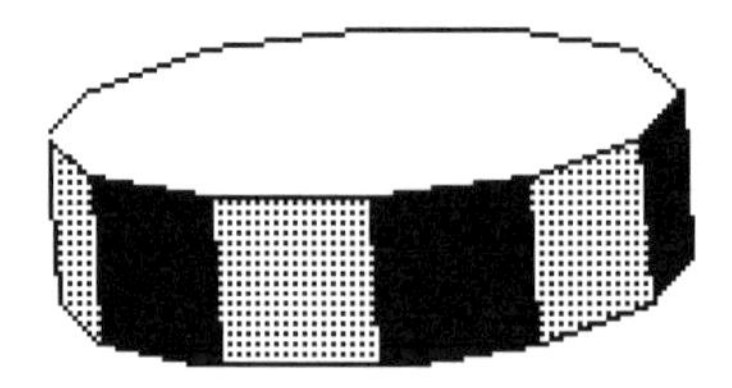

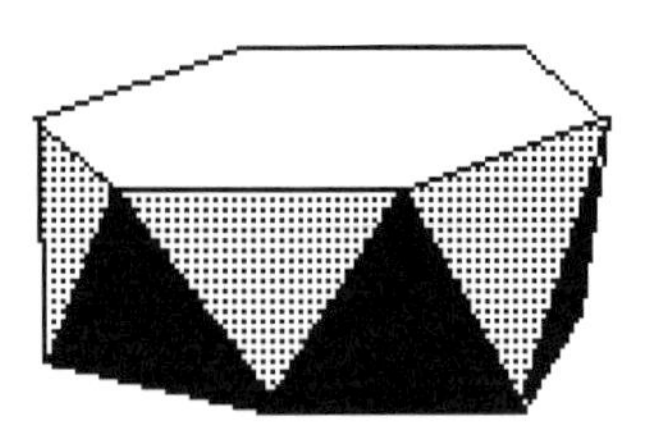

Platonic solids: If one slices off all of the corners from one of the platonic solids, each vertex can be replaced by a regular polygon. The lengths of the sides can be adjusted so that the remaining faces are also equilateral polygons. In fact, they are regular polygons, so the resulting polyhedron is semiregular.

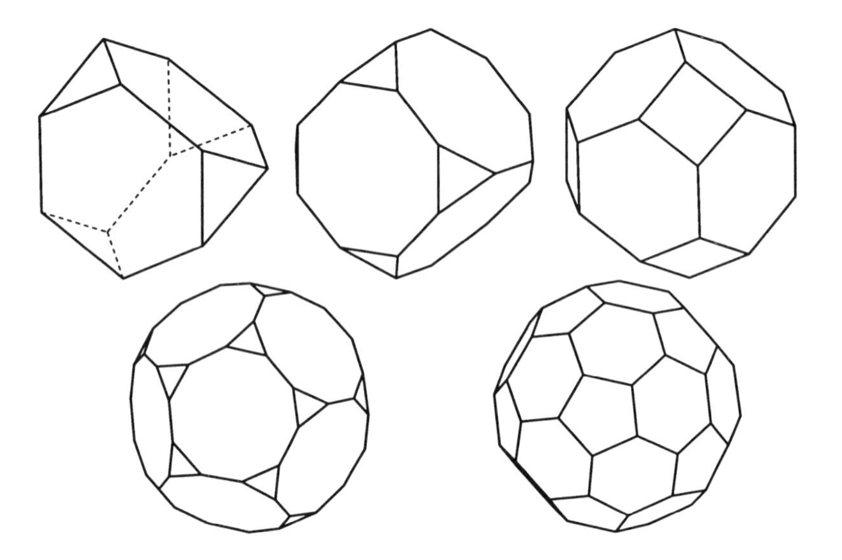

Among the five truncated polyhedra, the most familiar one is the truncated icosahedron, which is the standard design for a soccer ball. This particular polyhedron has recently gained prominence with the discovery of carbon molecules made of sixty atoms in this shape. They are known as "Buckyballs" or "Buckminsterfullerine" in honor of the architect and engineer R. Buckminster Fuller, who pioneered the use of polyhedra in architecture with structures known as *geodesic domes*.[1]

Starting with the cube, if we cut of larger corners so that the corresponding triangles touch at their corners, we get another semiregular polyhedron known as the *cuboctahedron*, which has vertex pattern

1. See H.W. Kroto ***et al***, "C_{60}: Buckminsterfullerene," *Nature* 318 (1985), 162–163.

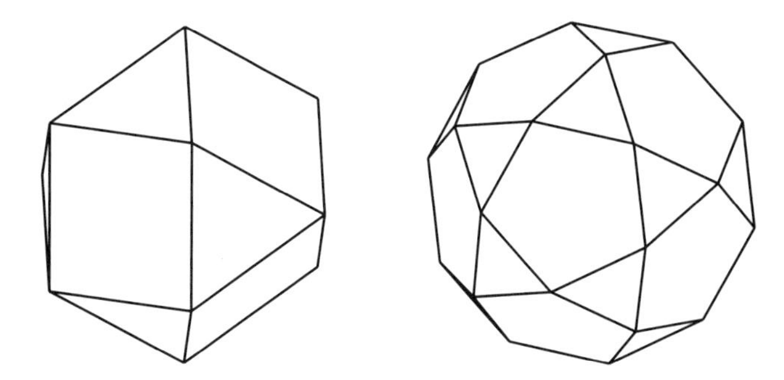

$\{4, 3, 4, 3\}$. Similarly, if we cut the corners off the dodecahedron we get the *icosidodecahedron*, which has vertex pattern $\{5, 3, 5, 3\}$.

The graphs of the great rhombicosidodecahedron and the great rhombicuboctahedron arise from those of the cuboctahedron and the icosidodecahedron by this same truncation process, but this time the faces are not all regular; the actual polyhedra have to be made by distorting the sides.

It turns out that there are two more vertex patterns with four polygons at each vertex. These are the pattern $\{4, 3, 5, 3\}$ and the pattern $\{4, 4, 4, 3\}$. The first corresponds to a polyhedron known as the *small rhombicosidodecahedron*. The second pattern has an interesting history. It corresponds to the polyhedron known as the *small rhombicuboctahedron*. Early in the twentieth century, a new semiregular polyhedron was discovered that has the same vertex pattern $\{4, 4, 4, 3\}$. (L.A. Lyusternik credits this discovery to V.G. Ashkinuz [24]; In [2], J.C.P. Miller is cited as the discoverer.) J. Malkevitch points out [31] that in fact, D.M.Y. Sommerville had Schlegel diagrams for both polyhedra in a paper in 1905. The pseudorhombicuboctahedron, as it is called, differs from the rhombicuboctahedron in an important way: The rotational symmetries of the figure do not take every vertex to every other vertex.

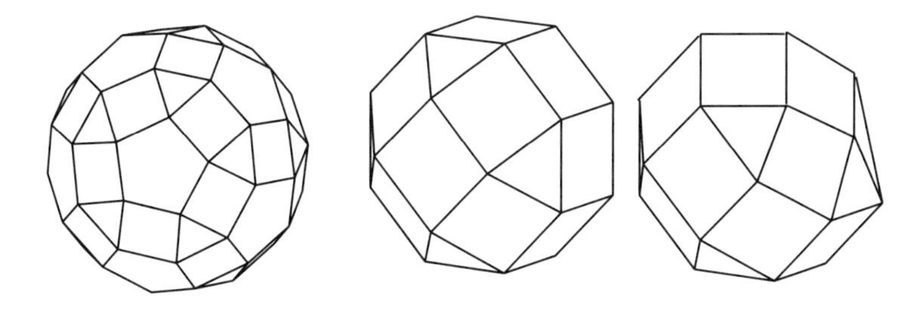

Problem

Draw Schegel diagrams for the rhombicuboctahedron and for the pseudorhombicuboctahedron.

Problem
How many isometries are there of the rhombicuboctahedron? How many are there of the pseudorhombicuboctahedron?

There are two more semiregular polyhedra, known as the *snub cube* and the *snub dodecahedron*. Each has five polygons at each vertex. These two, together with the eleven polyhedra described above, make up the *Archimedean solids*. Here are the Schlegel diagrams for these two solids.

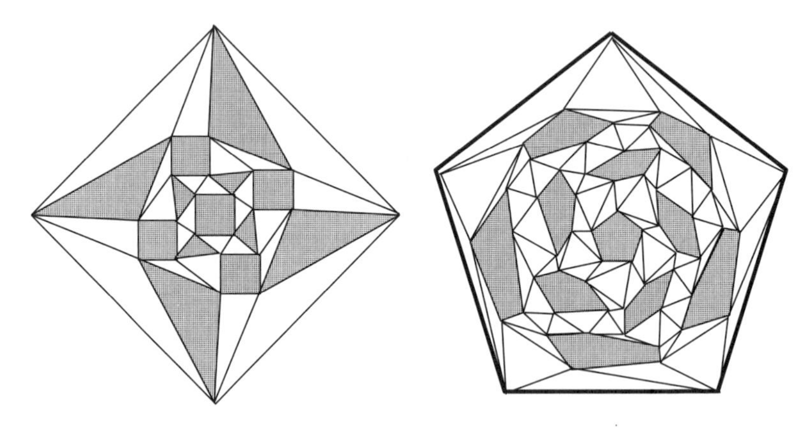

An important feature of these diagrams is *chirality*, the fact that they are different from their mirror images. Correspondingly, the polyhedra exist in two *enantiomorphous* forms. That is a fancy way of saying that there is a "left-handed" one and a "right-handed" one. These are the only Archimedean solids with this property. We can compare this with the semiregular tilings in the plane; you may recall that the (6, 3, 3, 3, 3) tiling also had two enantiomorphous forms.

The subject of mirror symmetry and handedness of objects is big enough to warrant a whole book, and in fact there is a fascinating one written by Martin Gardner [13]. There is a huge literature on polyhedra; an excellent starting place is the collection [31], where there are many more good bibliographic references.

4.4 Lines and Points: The Projective Plane and Its Cousin

The most obvious drawback of spherical geometry as an alternative to Euclidean geometry is the fact that straight lines meet in two points instead of one. There is another geometry, however, called *elliptic geometry*, which does not have this drawback. We will explore it briefly here.

In Section 4.1 we constructed a triangle ABC with two right angles, and its mirror image ABC', concluding from this that two lines met in two points (forming a "biangle"). There is another possibility, which at first blush seems unlikely: The points C and C' *might be the same point!*

Here is one model for this geometry. A **point** in the elliptic plane $\mathcal{E}$ will be defined to be an (unordered) *pair* of antipodal points on the unit sphere. A **line** in $\mathcal{E}$, which we will call an *e-line*, will be as before a great circle on the sphere.

Since the idea of using two points as one point is a bit confusing, there is another model that is a bit easier to use for some purposes. If we use stereographic projection to take the sphere to the plane, then the southern hemisphere is taken to the unit disc. Since every pair of antipodes has one point in the southern hemisphere (or both on the equator), we can represent points in our new geometry by points inside the unit disc together with pairs of antipodal points on the unit circle. In this model, straight lines are represented by straight lines through the origin and circle arcs that pass through two antipodal points on the unit circle. Angles are again measured in the Euclidean way.

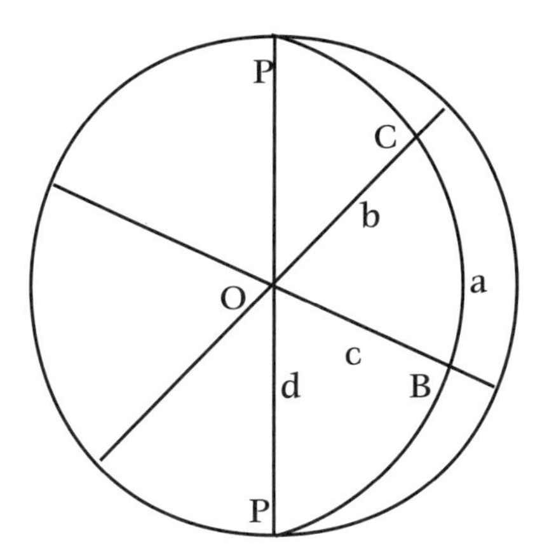

The e-lines a and b in the picture above meet at the point C. The e-lines a and d meet at the point P, which appears twice in the picture. It is easy to see from the picture that the angles in $\triangle OBC$ add up to more than $180°$. What is perhaps not so easy to see is that there is more than one choice for the triangle $\triangle OBC$! For instance, we can follow the straight line from O to B, then take the circle arc from B through P (jumping from one copy of P to the other) and on to C, and then take the line segment from C back to O. In fact, there are other choices of triangle (can you find them all?) with these same vertices.

The difficulty we are having stems from the fact that two points determine a line, but they do not determine a line *segment*. This is similar to the difficulty we had with the sphere, but there is more trouble ahead. It is no longer true that a line divides the plane into two pieces. In fact, if A and B are two points not lying on a line ℓ, then there is exactly one line segment joining A to B that does not meet ℓ. Since the entire straight line ℓ makes up a simple closed curve, this shows that the Jordan curve theorem is not true in $\mathcal{E}$.

The idea behind the Jordan curve theorem allows us to distinguish between triangles determined by three points. Of the two triangles OBC described above, the first one divides $\mathcal{E}$ into two regions; the second does not. How can we decide in general whether a polygon divides $\mathcal{E}$ or not? The answer comes from the relationship between spherical geometry and elliptic geometry.

There is a function $\Pi : S \longrightarrow \mathcal{E}$ that takes antipodal points in S to the single point they represent in $\mathcal{E}$. Using this function, we can draw pictures in the sphere that correspond to pictures in $\mathcal{E}$. For example, suppose we have a quadrilateral with vertices X_1, X_2, X_3, and X_4. In the sphere there are two points corresponding to each of these points. Pick a point Y_1 for which $\Pi(Y_1) = (X_1)$. The e-line determined by X_1 and X_2 corresponds to a great circle α in S through Y_1 and its antipode Z_1 containing the two points corresponding to X_2. If we choose an e-line segment in $\mathcal{E}$ joining X_1 to X_2, then the points taken to that e-line by Π make up two great circle arcs in α. But only one of them has Y_1 as an endpoint. The other endpoint is a point Y_2 corresponding to X_2. Its antipode Z_2 is the endpoint of the arc starting at Z_1. Proceeding in this way, we get a path joining Y_1 to Y_2 to Y_3 to Y_4 in the sphere. Now, the last step is to take an arc starting at Y_4 corresponding to the e-line segment from X_4 to X_1. The other end of this arc is either Y_1, or it is Z_1. We want the last point to be Y_1, so we will say that an *e-polygon* is one in which the corresponding curve in the sphere is a Jordan curve. Notice that if we had started at Z_1 instead of Y_1, the result would be to come back to Z_1 again, so our choice doesn't matter for our definition.

Problem

Show that an e-polygon is represented by a path in the unit disc that jumps across the unit circle boundary an even number of times. How many e-triangles are there joining three points A, B, and C?

We will define a tiling of $\mathcal{E}$ in the usual way, being careful, however, that we use only e-polygons. Now, suppose we have such a tiling. Using Π, we can then produce a tiling of the sphere, where each tile of $\mathcal{E}$ determines two tiles of S. Which tilings of the sphere correspond to tilings of $\mathcal{E}$? The answer comes from looking at the *antipodal map*. This is the isometry **A** of S that interchanges each point with its antipode. This can be achieved, for example, by the composition **FR**, where **R** is rotation by 180° around the north and south poles, and **F** is the reflection across the equator.

Proposition 4.4.1
A tiling of S comes from a tiling of $\mathcal{E}$ if and only if the antipodal map **A** *is a symmetry of the tiling.*

Problem
Find all regular and semiregular tilings of $\mathcal{E}$.

Problem
Show that the projective plane can be divided into six polygonal regions each of which has a border with each other region.

Problem
Prove the six color theorem: Any map on $\mathcal{E}$ can be colored using no more than six colors.

If we have a tiling of $\mathcal{E}$ with V vertices, E edges, and F polygons, then the corresponding tiling of S must have $2V$ vertices, $2E$ edges, and $2F$ faces. So the Euler formula for a tiling of $\mathcal{E}$ is

$$V - E + F = 1.$$

There is another way of viewing elliptic geometry. Instead of stereographic projection, let's look at *gnomonic projection*. We take a ray drawn from the center of the sphere (of radius 1) instead of the north pole. Each ray through a point in the southern hemisphere meets the plane tangent to the south pole in a point. We may think of this map as taking a pair of antipodal points to a single point in the plane. The pairs of antipodal points on the equator, however, give lines that do not meet the plane. We think of these as corresponding to *ideal points* at infinity. The plane together with these ideal points is called the *projective plane*.

The main virtue of gnomonic projection is that it takes great circles in the sphere to straight lines in the plane. In navigation, this is extremely

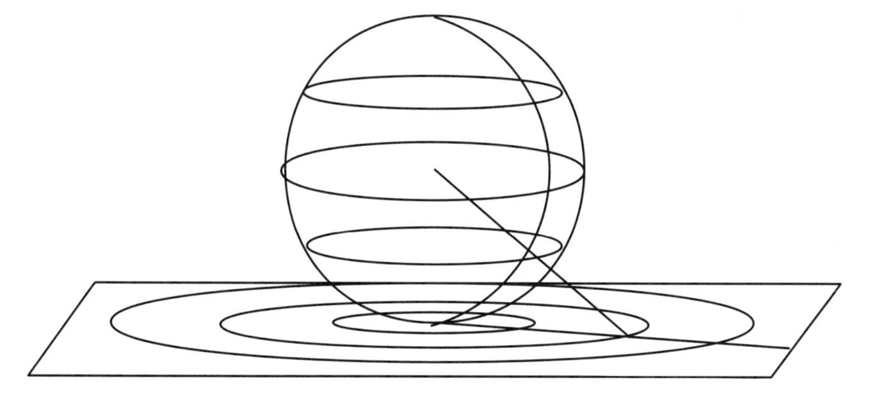

useful for course plotting. If we have a gnomonic projection of a piece of the globe, then we can connect two points on the map with a straight line to find the shortest flight path connecting these points. In practice, one can more easily follow a compass heading than a great circle arc. So the straight line in the gnomonic map is divided into segments P_1P_2, P_2P_3, etc. The longitude and latitude of each point P_i is determined. Then the compass heading needed to fly from point P_1 to point P_2, etc, is computed.

For our purpose, gnomonic projection allows us to think of the ordinary plane as a model for elliptic geometry. In this model, an ordinary Euclidean straight line is a straight line, and an ordinary point is a point. However, there is an additional ideal point added in each direction: Two lines that are parallel in the Euclidean sense are thought of as meeting at an ideal point. The collection of ideal points forms an ideal line. In this geometry, any two points determine a line, and any two lines intersect at one point.

Perhaps the most elegant feature of projective geometry is this symmetry—two points determining a line and two lines determining a point. In fact, theorems in projective geometry can be "dualized" by replacing point by line and line by point, giving rise to new, correct theorems. Using gnomonic projection, this turns out to be easy to explain.

A point in the projective plane can be thought of as a line through the origin in three-dimensional space. On the other hand, any line through the origin determines uniquely a plane through the origin that is perpendicular to that line. The plane through the origin determines a great circle by intersection with the unit sphere, or, by gnomonic projection, a straight line in the projective plane. So starting with a point in our plane, we can construct a line and vice-versa. Note that if we start with the origin, which corresponds to the north and south poles of the sphere, we end up with the ideal line. Conversely, if we start with an ideal point, we end up with a line through the origin.

Problem

What are the straight lines that correspond under this construction to points on the unit circle in the plane? (By unit circle, I mean the circle of Euclidean radius 1 around the point $(0, 0, -1)$ *in the plane* $z = -1$.*) Why is the unit circle special in this construction?*

Polar Coordinates. We will now look at the description of lines in the (Euclidean) plane. Let's denote by $\mathcal{M}$ the set of lines. Why $\mathcal{M}$? We will see the reason shortly. Since with one exception, every line in the projective plane corresponds via duality to a Euclidean line, and since every line corresponds to a point in the projective plane, we can think of the lines as making up the projective plane with one point removed. This useful notion allows us to make sense of the idea of two lines being "close to each other."

Let's pause for a moment to think about this. Suppose we plotted several points on a piece of graph paper and then wanted to pass a line through these points. Of course, if there are more than two points, the chances are that they won't exactly lie on any line. So we try to come close by drawing a line that almost passes through these points.

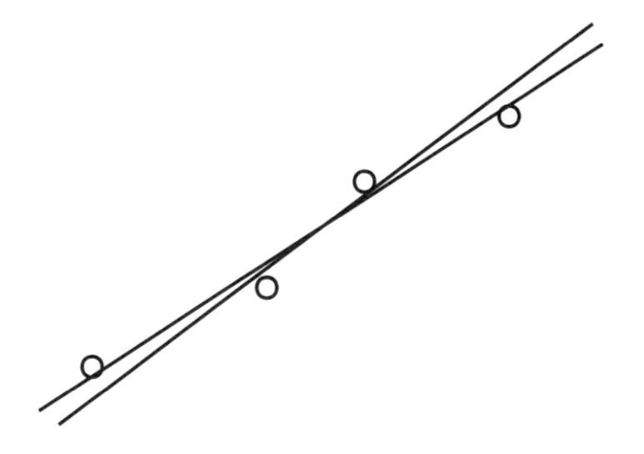

More than one such line may look pretty good. For example, two parallel lines may both look close, or two lines that intersect as in the picture above. If two different people tried to draw a "best" line, they would not necessarily draw the very same line. But we would expect the lines to look more or less the same. Intuitively, such lines ought to be close to each other. What is the property of two lines that makes us think they are close to each other?

Surely it is *not* where they intersect. Two parallel lines intersect way out at an ideal point, while we can tilt a line slightly at any point along it to get a line close to it that intersects at that point.

One property we can identify of close lines is that they are tilted at approximately the same angle. But that is not enough. Two parallel lines may be close together or far apart. Conversely, two lines making a very small angle always will be far apart if we go far away from the point of intersection. So we also want the lines to be close together *near some reference location* (for instance, near where we plotted our points).

With this in mind, let us now describe a coordinate system for lines in the plane, such that two lines that we want to consider to be close have close coordinates. First let us review the basic idea of coordinate geometry. To describe a point in the plane in coordinate form, it is necessary to (a) choose an origin, (b) pick a pair of coordinate axes through that origin (preferably orthogonal), and (c) mark off scales on the axes. Then every point P has a pair of coordinates (x, y) and vice versa. There is a correspondence between points and pairs of real numbers.

A straight line can then be described in various ways, using, for example, the principle that two points determine a line—this leads to the "two point formula":

$$\frac{y - y_0}{x - x_0} = \frac{y_1 - y_0}{x_1 - x_0}.$$

Is there a nice coordinate system for describing lines the way we describe points, so that a correspondence is set up between numbers and lines? The formula above is not very satisfactory. True, given numbers x_0, y_0, x_1, and y_1 we can determine a line, but this is a horribly redundant scheme. Another possibility is to use (A, B, C) to describe the line whose equation is $Ax + By + C = 0$. This still has redundancy (more about this later). The slope-intercept formula $y = mx + b$ doesn't have any redundancy, but not every line can be written in that form.

Here is a lovely scheme for describing lines that avoids these problems.[2] Envision a straight line as a (very long!) ship sailing in the ocean. Suppose a sonar operator is trying to locate the ship by bouncing a signal off its hull. The operator is located at the origin in the plane. If the operator bounces a signal perpendicularly off the hull of the ship (that is, off the line) it will bounce straight back to the operator. This will then give two numbers: an angle ϕ that indicates the direction of the signal and a distance r that the signal travels. Let us call these "reverse polar coordinates" (ϕ, r) for the line. (This is just to remind us that we are not doing the usual polar coordinates for points in the plane.) If we allow r to be equal to 0, then every line can be described by a pair of reverse polar coordinates. There is redundancy, but it is a mild sort of redundancy, since different nearby coordinates correspond to different lines. If we allow r to be positive, negative, or zero, then the redundancy can be described in a simple way: (ϕ, r) and $(\phi + 180, -r)$ describe the same line (and therefore also $(\phi + 360, r)$, $(\phi + 540, -r)$, etc). A representative coordinate pair for any line can be uniquely found in the infinite strip $\{0 \leq \phi < 180, -\infty < r < \infty\}$. We can include the edge of this strip, $\phi = 180$, together with the rule that every point $(180, r)$ corresponds to the point $(0, -r)$. In other words, the set of lines is described by an infinite ribbon with the edges glued together after twisting one edge relative to the other.

The fact that the ribbon is infinite is a bit of a nuisance. To fix this we have to use some trick like stereographic projection to compress the strip down to a finite size. We can then visualize the ribbon as a rectangle (with two opposite sides missing, since they represent points at infinity). The gluing process then produces that marvelous mathematical toy, the *Möbius band* $\mathcal{M}$.

---

2. I am grateful to the late Prof. Chih-Han Sah of SUNY Stony Brook for this idea.

r > 0
0 180
r = 0
r < 0

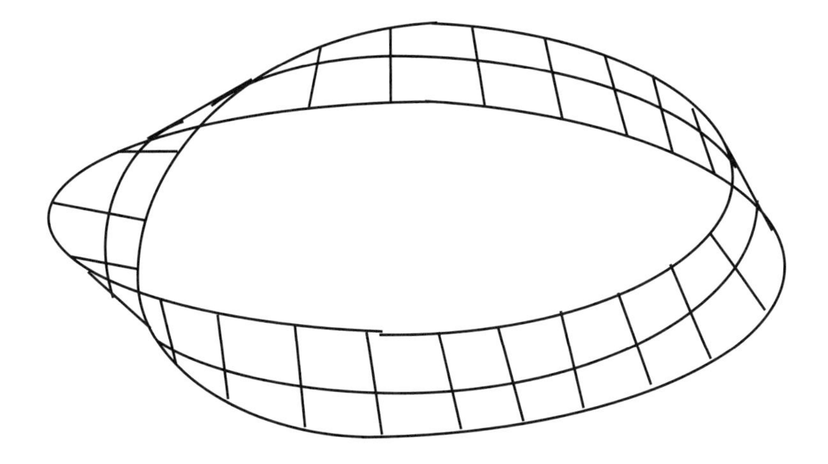

The centerline of $\mathcal{M}$ corresponds to the curve $r = 0$ (which becomes a circle after the ends have been glued together). This curve corresponds to the striaght lines that pass through the origin. In other words, $r = 0$ is the equation of a point! Remember that a "point" in $\mathcal{M}$ corresponds to a line in the plane. A point in the plane can be thought of as being described by the set of lines passing through that point, that is, by a *curve* in $\mathcal{M}$!

Problem

Find the equation of a point P in reverse polar coordinates. (This requires trigonometry, but only the definition of cosine.) Compare your answer with the equation of a line in polar coordinates in the plane.

If you have worked out the preceding problem correctly, you will see that the curve on $\mathcal{M}$ corresponding to the point P winds once around the band, and it crosses the equatorial curve at exactly one point. Why? Because there is one line through P that passes through the origin. More generally, the curves corresponding to points P and Q always cross at one point, because two points determine a line. The Möbius band has the property that any curve that goes once around the band must cross the centerline. This corresponds to the amazing fact that cutting along the centerline does not divide $\mathcal{M}$ into two pieces. (If you have never done this, by all means build a Möbius band out of paper and try it!)

Problem

Use this property of $\mathcal{M}$ to say something about moving a line around in the plane without letting it hit the origin so that it rotates through $180°$ *and comes back to itself. What about rotating through* $180k°$?

Remember that using duality, we can think of the lines in the plane as corresponding to points in the projective plane. If $D : \mathcal{M} \longrightarrow \mathcal{E}$ is the function that assigns to each line in the plane the dual point in the projective plane, then D takes $\mathcal{M}$ onto every point except the one corresponding to the line at infinity. If we visualize $\mathcal{E}$ as the unit disc, then the point that is dual to the line at infinity is the center of the disc. The centerline of the Möbius band, as we saw, consists of the lines through the origin. D takes these lines to (pairs of) points on the boundary of the unit disc. So the projective plane and the Möbius band are cousins; if we remove the center of the unit disc and then glue opposite points on the boundary together, we have the Möbius band.

Actually, this is physically impossible, for a couple of reasons. First of all, the edge of the Möbius band is not supposed to exist. And the points on this nonexistent edge correspond to the center of the disc. If instead of removing just the center we cut a small hole in the center of the disc, then in theory we could glue opposite points on the unit circle together and make a Möbius band. But to physically achieve this, we would need to have the disc made of some stretchable material. The tiny hole in the center would end up getting streched a lot. This process of stretching an object without allowing it to rip apart is a basic process in *topology*, which Kasner and Newman refer to as "Rubber-sheet Geometry" in their classic book about mathematics [22]. Here is the relevant fact from topology, which is proved using the function D:

Proposition 4.4.2

The Möbius band is topologically equivalent to the projective plane with one point removed.

It is a fact, rather difficult to prove, that it is impossible to build a model for the entire projective plane with no self-intersections. The Möbius band, which contains all but one point of the projective plane, is as close as we can come!

5 CHAPTER More Geometry of the Sphere

5.1 Convex Polyhedra Are Rigid: Cauchy's Theorem

We saw in the last chapter that a tessellation of the sphere by regular polygons determines a polyhedron with regular faces inscribed in the sphere. However, it is certainly possible to construct polyhedra with regular faces that are not inscribed in the sphere. In our classification of regular and semiregular polyhedra, we saw that Euler's formula severely limited the possible polyhedra that we could construct. By analysis of the numerical relations implied by the Euler formula and the polygonal faces, we were able to find all possible candidates.

One important question that we ignored was this: Is the combinatorial data enough to completely describe the polyhedron? In other words, suppose two people attempt to assemble the same polyhedron. Each is given an identical collection of polygons together with assembly rules indicating which polygon is to be attached to which along a common edge. Each assembles the polyhedron according to these rules. Will the resulting polyhedra be identical (that is, congruent)?

The example of the pseudo-rhombicuboctahedron serves as a warning. It shows that knowing only what each vertex looks like is not sufficient. So let us assume that we have been more precise: We have specified exactly which edge of each polygon corresponds to which edge of another polygon. For example, we could have a diagram that is the Schlegel diagram for the proposed polyhedron. (Remember that the rhombicuboctahedron and its sibling had different Schlegel diagrams.) Suppose we have specified the dimensions of each polygonal face in such a way that corresponding edges of adjacent faces have the same lengths. Does this determine the polyhedron uniquely?

As stated, this question turns out to be very difficult to answer. Here is a simple example. Suppose we want to build a tetrahedron. We will be gluing together four triangles along their edges. The Schlegel diagram is a very simple one. If we specify the lengths of the six edges of the diagram, can we build the resulting tetrahedron? Well of course, we need to know that each face is an honest triangle. This means that no side can be longer than the sum of the other two sides. But this is definitely not enough, as the following example shows.

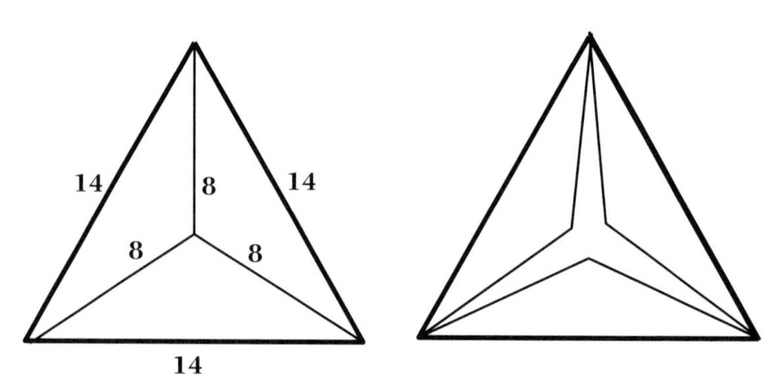

Suppose we take an equilateral triangle with sides of length 14 and three isosceles triangles with sides of lengths 8, 8, and 14. The Schlegel diagram (left) shows how we plan to glue the polygons together. But after gluing the long sides of the isosceles triangles to the equilateral triangle, it is no longer possible to glue the remaining edges together. They don't reach far enough.

This example is bit too simple. In fact, the large angles of the isosceles triangles are bigger than 120°, which means that we can't even fit these three triangles together. But it will suffice as a warning: Knowing that edges match up in length is not enough to insure that we can assemble the polyhedron. So let's assume that it is possible to assemble the polygons into a polyhedron. Will there be only one way to do it?

The answer is: not necessarily. For example, start with our friend the small rhombicuboctahedron. There is a plane that passes through eight vertices of it, cutting it into two pieces. We got the pseudo-rhombicuboctahedron by rotating one of the pieces and reattaching it. This changed the

Schlegel diagram. Suppose instead we *reflect* the smaller piece through the plane and reattach it. We get another polyhedron with the same Schlegel diagram as the small rhombicuboctahedron, but it is a very different figure.

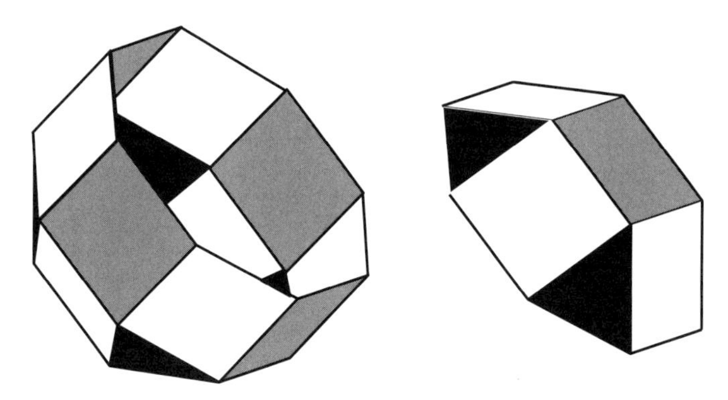

The obvious difference between the newly constructed polyhedron and other polyhedra is that part of it is "caved in." In this section we will avoid such polyhedra by concentrating on geometric objects that are *convex*. A geometric object in the plane or in three-dimensional Euclidean space is convex if whenever it contains points A and B, it contains the entire line segment AB joining them. A triangle in the plane is always convex, but not all four-sided polygons are convex.

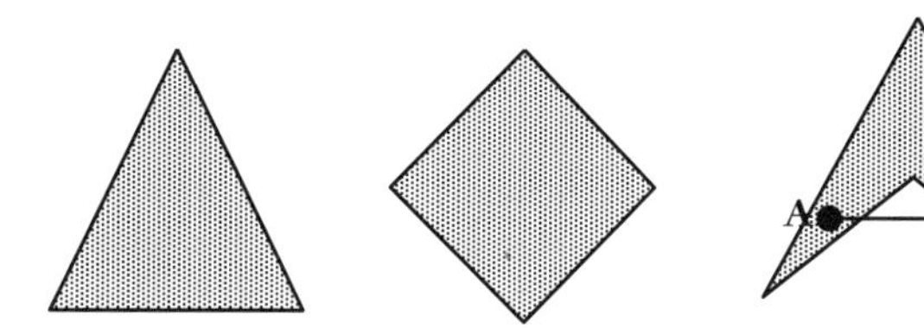

If X is a convex object, then it divides space up into three kinds of points. The points outside the object are called *exterior points*. To be more precise, an exterior point must have the property that some small disc or ball around the point consists entirely of points outside of X. A convex set is called *closed* if all points not in X are exterior points. This rules out, for example, the possibility of taking a triangle and throwing out one if its vertices. It is easy to check that such a set is convex, but the missing vertex is not exterior to it. From now on all convex objects we consider will be closed.

Points in the object are divided up into *interior points* and *boundary points* . The difference between them is that if A is an interior point, then all the points within a small disc or ball around the point are also in X. A convex set X is *bounded* if it is entirely contained in some large disc or ball. Not all convex sets are bounded; for instance, a half plane is an unbounded convex set.

Now suppose X is a bounded (closed) convex object in the plane. Then the boundary points always turn out to form a simple closed curve. We will call such a curve a *convex curve*. A convex curve made up of straight line segments is a *convex polygon*. In three dimensions, the boundary points of a convex object make up a *convex surface*, provided that the X does not lie entirely in a plane. If the surface is made up of polygons, it is called a *convex polyhedron* .

Now that we have a clean and precise definition of the geometric objects we want to study, we can formulate the theorem that is the goal of this section. It was stated and proved by the great French mathematician Augustin Louis Cauchy in 1813 (see [24], p. 66–81, or [6], p. 226–231 for a thorough treatment of the theorem).

Theorem 5.1.1 (Cauchy's Theorem) *Two convex polyhedra comprising the same number of equal similarly placed faces are superposable or symmetric.*

A consequence of Cauchy's theorem is that convex polyhedral surfaces are *rigid*, meaning that they hold their shape. For some polyhedra, such as the cube, this is obvious. To see why, let us first look at polygons in the plane. Suppose we are given a collection of bars, each made out of unbendable, unstretchable material. They are connected together using ball joints, which allow the angles between successive bars to change, to form a planar polygon. Assuming that the bars remain in a fixed plane, when can the polygon change its shape?

If there are only three bars, forming a triangle, then the resulting polygon must hold its shape; this is a consequence of the SSS congruence theorem in Euclidean geometry. If there are more than three bars, then it is pretty clear that such polygons are never rigid. With four bars we can always "flex" the polygon, making two angles smaller and the other two larger.

If the quadrilateral is convex, then the four angles alternately increase and decrease in size. (The angle at vertex A in the picture below decreases, while B increases, when we read from left to right.) the nonconvex case,

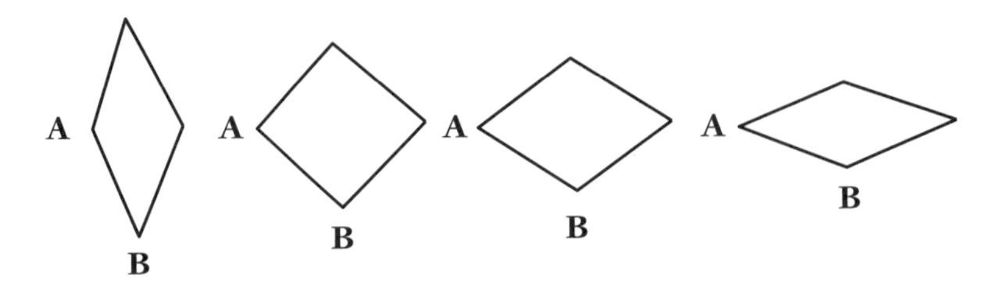

this is not necessarily true. The angles at vertex A increase along with those at B in the picture below.

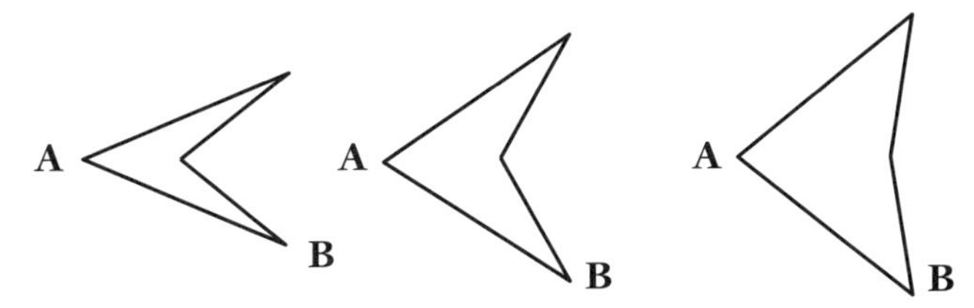

Imagine a small sphere centered at one vertex of a cube. The intersection with the sphere of each square adjacent to this vertex is a 90° arc of a great circle on the sphere. The complete intersection of the cube with the sphere is a spherical triangle. If the three squares that come together at one corner of the cube could flex, then likewise, the triangle on the sphere would have to flex. (Notice that the length of each side cannot change; it depends only on the angle of the polygonal face.) But triangles in the sphere are rigid, by SSS (which still works on the sphere). Consequently, three squares can be glued together at a corner in only one way.

The same statement is definitely not true for an octahedron. The four triangles that come together at one vertex form a flexible structure. If you build a model of such an object (a square pyramid without its base), you can quickly discover that it is flexible. The argument above shows that flexing one corner of an octahedron corresponds to flexing a spherical quadrilateral. Again two opposite angles get larger and the other two get smaller. In the pyramid this translates into the observation that two of the *dihedral angles* between adjacent triangles get larger and two get smaller.

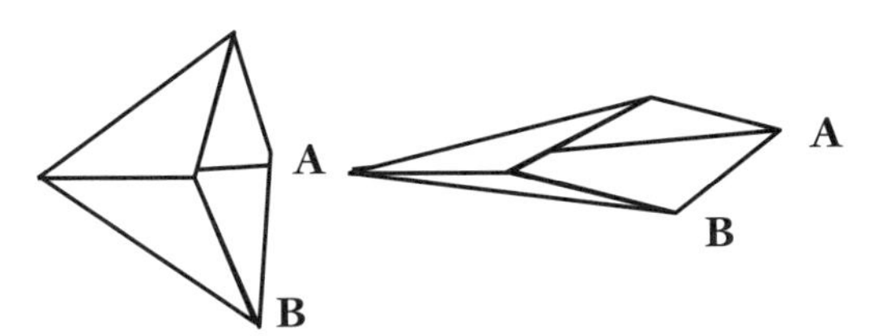

Yet while half of an octahedron is a flexible object, the entire octahedron has no flexibility at all. The easiest way to understand this is to notice that as we flex one corner of the octahedron as shown above, the dihedral angle along the edge from that corner to the vertex A decreases, while the angle from the corner to the vertex B increases. Now, whichever way the dihedral angle along the edge AB is supposed to change, it will be inconsistent with this pattern of alternate dihedral angles at a vertex changing in opposite directions.

This argument is good enough to explain why any convex octahedron is rigid, but it is a bit too special to take care of more general polyhedra.

Cauchy's proof generalized this argument in two ways. First, he stated and proved a general lemma about the ways in which convex planar or spherical polygons flex. Then he applied Euler's theorem to show that the ways that the dihedral angles in a polyhedron would have to change around each vertex was always inconsistent.

Lemma 5.1.2

[Cauchy's Lemma] Suppose we transform a convex (spherical or planar) polygon $A_1A_2 \cdots A_n$ into another convex polygon $A_1'A_2' \cdots A_n'$ in such a way that the lengths of the sides A_iA_{i+1} remain unchanged. If the angles at the vertices $A_2, A_3, \ldots, A_{n-1}$ remain unchanged or increase, then the length of the remaining side A_nA_1 must also increase.

When $n = 3$, this is the "caliper lemma" from Chapter 1. Cauchy attempted to prove this lemma by an induction argument on the number of sides in the polygon. His strategy was to increase the angles one at a time, keeping the other angles fixed. Unfortunately, this doesn't always work; Cauchy missed the possibility that in increasing the angles one at a time, there might be an intermediate stage in which the polygon was not convex.

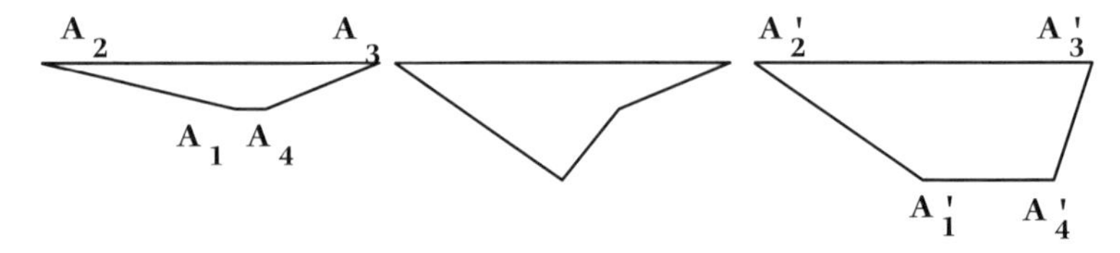

A corrected proof was provided by Steinitz in 1934. Since then there have been various correct proofs of this lemma, of which perhaps the simplest is due to Schoenberg and Zaremba.[1] We will look at their proof of the planar version of the lemma. The spherical case (which is actually the case we need for Cauchy's theorem) is not too much more complicated.

PROOF of Cauchy's Lemma

Among the vertices of the original polygon, choose the vertex A_k farthest from the line containing A_n and A_1. (If there is more than one, there is a similar argument.) Picture the polygon as having A_1A_n as part of the x–axis, with the x coordinate of A_n larger than that of A_1, and A_k as being on the positive y–axis.

Now construct the altered polygon in such away that A_k does not move and no vertex of the new polygon is higher up on the y axis. Let (x_i, y_i) be the coordinates of the point A_i. Then the length of A_1A_n is given by

1. I.J. Schoenberg and S.K. Zaremba, On Cauchy's lemma concerning convex polygons, *Canad. J. Math.* **19** (1967), 1062–1077.

$$x_n - x_1 = (x_2 - x_1) + (x_3 - x_2) + \ldots + (x_n - x_{n-1}).$$

Some of these quantities are positive and some are negative. All we need to see is that each one of them gets larger when we move to the new polygon; it will then follow that the length of $A_1'A_n'$, which is at least $x_n' - x_1'$, will be larger than that of A_1A_n.

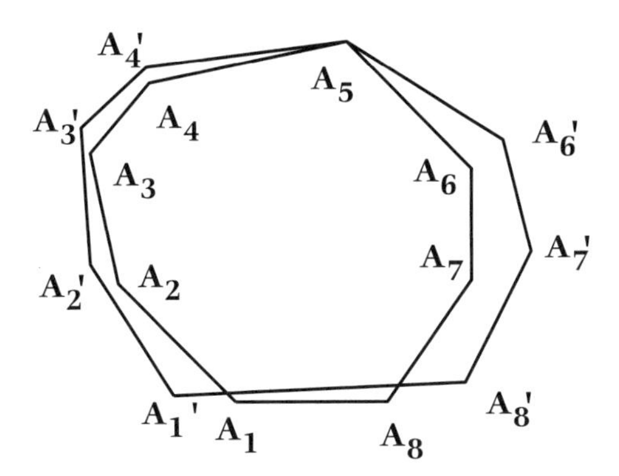

This is easy to see geometrically. The edges to the left of the vertex A_k must rotate counterclockwise, while those to the right must rotate clockwise. In the picture, for example, $x_4 - x_3$ is positive, and after the segment A_3A_4 is rotated, it increases; $x_2 - x_1$ is negative, and after rotation it also increases (that is, gets less negative.) The edges on the other side of A_5 are rotated in the opposite direction, with the same consequence.

Corollary 5.1.3
Suppose a convex polygon is transformed into another convex polygon without changing the lengths of its sides. Mark each vertex whose angle increases with a + sign and each which decreases with a − sign. Then in making a circuit of the vertices in order, we must find at least two changes of sign from plus to minus and two from minus to plus (or no changes at all.)

Proof
Suppose some vertex angle increases. Then obviously, some other angle decreases, so the number of sign changes is at least 2. Also obviously, it is an even number. So we only need to rule out the possibility of exactly two changes of sign. But if there were exactly two, then we could divide the polygon into two pieces by drawing a diagonal connecting the midpoints of two sides, in such a way that all the angles on one side of the diagonal increase and all the angles on the other side decrease. But then by the Cauchy lemma the diagonal would have to get bigger and also smaller. ∎

PROOF of Cauchy's Theorem

Suppose we have two convex polyhedra X and Y with corresponding faces that are congruent. To decide whether the polyhedra themselves are congruent, we look at each edge and see whether the dihedral angle (The angle between the faces sharing that edge) has changed. If none of the dihedral angles is different, then it is pretty easy to see that the X and Y are congruent.

Take the Schlegel diagram of the polyhedron. Recall that this is a graph with one vertex for each vertex of the polyhedron and one edge for each edge of the polyhedron. We can picture this as a graph in the plane, and Euler's formula holds for it. Now let us put a label on some edges of the diagram, as follows: If the dihedral angle at the edge in Y is larger than the one in X, put a $+$ sign. If the angle is smaller, put a $-$ sign. If they are the same, then leave it unmarked.

Next, delete from the diagram all unmarked edges. What remains is a graph G that has each edge marked with a $+$ or $-$. This graph has V vertices, E edges, and F regions, and $V - E + F = 1 + C \geq 2$, where C is the number of components of G. (Since G need not be the graph of a polyhedron, we don't know the value of C.)

Now look at the edges that emanate from one vertex V of the graph G. As we go once around V reading the markings on the edges, we count the number of sign changes. By the Corollary to the Cauchy lemma, there must be at least four such changes around V. Let S be the total number of sign changes we find as we go around every vertex Then

$$S \geq 4V. \tag{5.1.1}$$

On the other hand, as we walk around any region determined by G, we can also count changes of sign as we go from one edge to the next. If the region has three sides, there can be at most two sign changes (since it has to be an even number), while if the region has four sides, there can be at most four sign changes, etc. If F_n is the number of regions with n sides, then we have the equations

$$F = F_3 + F_4 + F_5 + F_6 + \cdots, \tag{5.1.2}$$

$$2E = 3F_3 + 4F_4 + 5F_5 + 6F_6 + \cdots \tag{5.1.3}$$

and

$$4V \leq S \leq 2F_3 + 4F_4 + 4F_5 + 6F_6 + \cdots. \tag{5.1.4}$$

Now, by Euler's formula, $4V - 8 \geq 4E - 4F$, so putting 5.1.2 and 5.1.3 together,

$$4V - 8 \geq 2F_3 + 4F_4 + 6F_5 + 8F_6 + \cdots. \tag{5.1.5}$$

Now subtract equation 1.5.4 from equation 5.1.5 to get

$$-8 \geq 2F_5 + 2F_6 + 4F_7 + 4F_8 + 6F_9 + \cdots.$$

This equation is impossible. So the graph G does not exist! In other words, there are no labeled edges at all. That proves the theorem.

The assumption of convexity was crucial in this proof. Cauchy's lemma is not true for nonconvex polygons. Furthermore, we know it is possible to build a nonconvex polyhedron with the same polygons as a convex polyhedron. However, we could not bend one of the polyedra into the other without breaking it apart. Of course, half of an octahedron is flexible. To speak about rigidity in a more general setting, we should first define a *closed polyhedral surface* to be a polyhedron in which each edge is shared by exactly two faces. The *Rigidity Conjecture* can then be stated as follows: a closed polyhedral surface cannot be deformed without stretching or bending the faces or breaking it apart. This conjecture was apparently formulated by Euler, and it remained an unsolved problem until 1978, when Robert Connelly found a counterexample! Instructions for building a flexible polyhedron can be found in Connelly's article, The rigidity of polyhedral surfaces, *Math. Mag.* **52** (1979), 275–283. A complete survey of the subject can be found in [6].

5.2 Hamilton, Quaternions, and Rotating the Sphere

In studying the isometries of the plane, and later in studying the isometries of the hyperbolic plane, we found a powerful computational tool in the complex numbers. With their help we could determine, for example, what the composition of two rotations around different points was. We now want to investigate the isometries of the sphere. The positively oriented isometries (see Section 3.3 for a discussion of orientation) turn out to be rotations; this is a theorem due to Euler.[2] Pick a pair of antipodal points on the sphere, or take the line a through these two points in 3-space. Then a rotation $\mathbf{R}$ through an angle θ around this line is an isometry of the sphere. Now suppose we take a second pair of antipodes and the corresponding line b and take a rotation $\mathbf{R'}$ around b through an angle ϕ. What is the result of taking the composition $\mathbf{R'R}$?

This is not an easy question to answer. It turns out to be a rotation about an axis c through an angle ψ. But what axis? What angle? This

2. L. Euler, Formulae generales pro translatione quacunque corporum rigidorum, *Novi Comm. Acad. Sci. Imp. Petrop.* **20** (1775), 189–207.

problem was solved by Olinde Rodrigues in 1840.[3] Not long afterward (October 1843) William Rowan Hamilton discovered the *quaternions*, a higher-dimensional generalization of the complex numbers. Using these numbers he was able to describe the rotations of the sphere. [4] We will examine quaternions in this section. For a full treatment of the history of the problem of rotations of the sphere, see [1], chapter 1).

We begin by defining a new imaginary number j, which has the property $j^2 = -1$ in common with i. A quaternion q is a "number" of the form

$$q = \alpha + \beta j,$$

where $\alpha = a + bi$ and $\beta = c + di$ are complex numbers. We will try to imitate the complex numbers by defining the arithmetical operations. If $q' = \alpha' + \beta' j$ is another quaternion, then the sum $q + q' = (\alpha + \alpha') + (\beta + \beta')j$ is easy to describe. Before we determine the rule for multiplication, let us think about the rules of arithmetic. The *distributive law* of arithmetic decrees that $(x + y)z = xz + yz$ and $x(y + z) = xy = xz$. So in particular, $(c + di)j = cj + (di)j$. If the *associative law* of multiplication holds, then $(di)j = d(ij) = dk$, where k is defined by the formula

$$k = ij.$$

Now we may rewrite a quaternion q by

$$q = (a + bi) + (c + di)j = a + bi + cj + dk.$$

The question is, what is the number k? If the *commutative law* for multiplication holds, then we have

$$k^2 = k(ij) = (ki)j = (ik)j = (i(ij))j = ((ii)j)j = (-j)j = +1.$$

This seems perfectly reasonable. But notice that then

$$(1 + k)(1 - k) = 1 - k + k - k^2 = 1 - 1 = 0.$$

This is not good news. If the product of two numbers is 0, then one of them had better be 0. If not, division becomes a problem. (If we are allowed to divide by any number except 0, then we can first divide by one factor and then by the other, with the result that we can divide by 0!)

So either $k = 1$ or $k = -1$. But if $k = 1$, then $i(i + j) = i^2 + k = 0$, while if $k = -1$, $i(i - j) = 0$. Now the same reasoning as before would say that

3. O. Rodrigues, Des lois géométriques qui régissent les déplacements d'un système solide dans l'espace, et de la variation des coordonnées provenant de ses déplacements considérés indépendamment des causes qui peuvent les produire, *J. de Mathématiques Pures et Appliquées* **5** (1840), 380–440.
4. W.R. Hamilton, On quaternions; or a new system of imaginaries in algebra, *Phil. Mag.*, 3rd. ser., **25** (1844), 489–495.

$i = j$ or $i = -j$; in other words, our new numbers would just be complex numbers!

Hamilton wrestled with this problem for a long time. Originally, he sought a multiplication rule for triples $a + bi + cj$ of numbers. After much struggle, he was led to the surprising realization that he needed to abandon the commutative law of multiplication; he decided on the rule

$$ij = k = -ji. \tag{5.2.1}$$

Hamilton's struggle to discover the multiplication rules of quaternions is described in a lovely article by B.L. van der Waerden, "Hamilton's Discovery of Quaternions,"(*Mathematics Magazine* **49** (1976), 227–234.) After deciding on the formula 5.2.1, van der Waerden reports, Hamilton was then led to the necessity of dealing with four-dimensional numbers, a difficult leap of insight since the concept of a fourth dimension seemed paradoxical at that time.

We may now describe the laws of arithmetic for quaternions. All the rules of arithmetic, including associative laws, the distributive law, and commutative laws hold, with the exception of the commutative law of multiplication. This last law is supplemented by equation 5.2.1. A consequence of this is the general rule of multiplication

$$qq' = (\alpha + \beta j)(\alpha' + \beta' j) = (\alpha\alpha' - \beta\overline{\beta'}) + (\alpha\beta' + \beta\overline{\alpha'})j. \tag{5.2.2}$$

Problem

Verify equation 5.2.2 by checking that it follows from the special case

$$j\alpha = \overline{\alpha}j.$$

What is the formula for the product of q and q' in terms of the real numbers a, b, c, d, etc.?

The *absolute value* of a quaternion q, denoted by $|q|$, is given by

$$|q|^2 = |\alpha|^2 + |\beta|^2 = a^2 + b^2 + c^2 + d^2.$$

Geometrically, it can be thought of as the (four-dimensional!) length of the line segment from the origin in 4-space to the point with coordinates (a, b, c, d). This is the generalization of the Pythagorean theorem to four dimensions.

The *quaternionic conjugate* q^* of the quaternion q is given by

$$q^* = a - bi - cj - dk = \overline{\alpha} - \beta j.$$

If $\beta = 0$, so that q is an ordinary complex number, q^* is just its complex conjugate. This definition is motivated by the following important fact, which can be checked by computation:

Proposition 5.2.1
$qq^* = q^*q = |q|^2$. If q and q' are quaternions, then $(qq')^* = q'^*q^*$. (Note that the order of multiplication is reversed.)

Corollary 5.2.2
For any two quaternions q and q', $|qq'| = |q||q'|$.

Proof
$|qq'|^2 = (qq')(qq')^* = (qq')q'^*q^* = q(q'q'^*)q^* = q|q'|^2q^* = qq^*|q'|^2 = |q|^2|q'|^2$. ■

Notice in the proof how convenient it is that * reverses the order of multiplication. Because multiplication is no longer commutative, that reversal was really important in making the argument work.

Problem
Let $q_1 = 1 + i - j$ and $q_2 = 2 + 2j + k$.

1. *Compute $q_1 + q_2$, q_1q_2, and q_2q_1.*
2. *Verify that $|q_1q_2| = |q_1||q_2|$.*
3. *Let $x = -\frac{1}{3}i + \frac{5}{3}j - \frac{1}{3}k$. Verify that $q_1x = q_2$.*
4. *compute xq_1.*

You may recall (from Chapter 2) that complex conjugation was the key tool in defining division. The same idea works for quaternions, but we must decide what we actually mean by division. First consider the problem of dividing 1 by q. If q is a quaternion (not 0), the quaternion $\frac{1}{q}$ should have the property that $q \times \frac{1}{q} = 1 = \frac{1}{q} \times q$. As we will see, this notation is misleading, so we will use the expression q^{-1} instead of $\frac{1}{q}$. By the same principle we used for complex numbers, we can quickly see that the formula

$$q^{-1} = \frac{1}{|q|^2}q^* \tag{5.2.3}$$

works. We need to use the fact that $qq^* = q^*q$ and the fact that multiplication by real numbers can be performed on either side to see that $q^{-1}q = 1 = qq^{-1}$. (Check this computation for yourself.)

Now the general problem of division of a by b can be formulated in two ways:

- Solve the equation $bx = a$ for x;
- Solve the equation $yb = a$ for y.

In general, these problems have *different* solutions.

Proposition 5.2.3
If $b \neq 0$, then $x = b^{-1}a$ solves the equation $bx = a$, while $y = ab^{-1}$ solves $yb = a$.

We may say that x arises by "dividing on the left," while y arises by "dividing on the right." If a or b is real, then $x = y$.

Problem

1. *When, if ever, can the equation $bx = a$ have more than one solution?*
2. *Find x and y such that $(j + k)x = y(j + k) = i$.*
3. *Suppose $(j + k)x = x(j + k) = q$. What possible values can q have?*
4. *Suppose $(j + k)x = -x(j + k) = q$. What possible values can q have?*

Complex numbers of absolute value 1 rotate the complex plane by multiplication. We can imitate this construction using quaternions, but there are two complications. The first is that quaternions make up a four-dimensional space, and we want to study rotations in three-dimensional space. The second is that because the commutative law fails for multiplication, there are two different ways to define multiplication by a unit quaternion. This fact turns out to be the key to understanding rotations in three dimensions.

First some notation. The symbol $\mathcal{H}$ (from the name Hamilton) denotes the four-dimensional space of quaternions. The *unit quaternions*, that is, the quaternions that have absolute value 1, form the *unit 3-sphere* S^3 in $\mathcal{H}$. (This is the generalization to one higher dimension of the unit sphere S in 3-space R^3.

A quaternion $q = a + bi + cj + dk$ can be thought of as the sum of a real number a and a *pure quaternion* $V = bi + cj + dk$. Hamilton called a a *scalar* and V a *vector*. V is the *vector part* of q, while a is the *scalar part*. We will think of R^3 as being the space of vectors, where $bi + cj + dk$ corresponds to the point with coordinates (b, c, d).

With this understanding, the vector i corresponds to the point $(1, 0, 0)$, or the unit vector along the positive x-axis in R^3. Likewise, j and k are the unit vectors along the positive y-axis and z-axis, respectively. (This is actually the standard notation used for these vectors in vector calculus today, although it is not usually mentioned that the letter i stands for the complex number!)

If $V = bi + cj + dk$ and $W = b'i + c'j + d'k$ are two pure quaternions (that is, vectors), then their product VW is not generally a vector. If we write the product as $x + X$, where x is the scalar part and X is the vector part, then we call $-x$ the *scalar product* and X the *vector product*. The vector product is usually denoted by $V \times W$; it is also called the *cross product*. The scalar product is usually denoted by $V \cdot W$; it is also called the *dot product*. These are both important operations in the study of vector calculus, but we will not be using them here.

Problem

Compute $V \times W$ and $V \cdot W$. Verify that $V \times W = -W \times V$. What can be said in general about $V \times V$?

In order to use quaternions to describe rotations in R^3, we need to solve the problem of recognizing when a quaternion is a vector; in other words, when the real part of a quaternion is 0. This has an elegant solution:

Proposition 5.2.4

A quaternion q is a vector if and only if q^2 is a nonpositive real number.

Proof

If $q = a + bi + cj + dk$, then using the rules for multiplying quaternions together, we compute $q^2 = (a^2 - b^2 - c^2 - d^2) + 2abi + 2acj + 2adk$. Now, if $a = 0$, then this reduces to $-b^2 - c^2 - d^2$, which is a negative real number or zero. On the other hand, if q^2 is a real number, then either $a = 0$ or b, c, and d all have to be zero. So either q is a vector or a scalar. Of course, if it is a scalar, its square can't be negative, so q must be a vector. ■

If $q = a+bi+cj+dk$ is a unit quaternion, then of course $a^2+b^2+c^2+d^2 = 1$. So we can find an angle θ with $0 \leq \theta \leq 180°$ and $a = \cos\theta$. Then we can write $q = \cos\theta + \sin\theta V$, where V is a vector of length one. For reasons to be explained in a little while, let us replace θ by $\frac{1}{2}\phi$, where now ϕ is an angle between 0 and 360°.

Define the transformation $\mathbf{L}_q : \mathcal{H} \longrightarrow \mathcal{H}$ by

$$\mathbf{L}_q(x) = qx.$$

Similarly, define the transformation $\mathbf{R}_q : \mathcal{H} \longrightarrow \mathcal{H}$ by

$$\mathbf{R}_q(x) = xq.$$

$\mathbf{L}_q$ is a *linear* function; that is, $\mathbf{L}_q(x+y) = \mathbf{L}_q(x) + \mathbf{L}_q(y)$. This is just the distributive law for multiplication over addition. Now if x and y are any two points in $\mathcal{H}$, then the line segment joining them has length $|y - x|$. The line segment joining $\mathbf{L}_q(x)$ to $\mathbf{L}_q(y)$ has length

$$|\mathbf{L}_q(y) - \mathbf{L}_q(x)| = |qy - qx| = |q(y - x)| = |q||y - x| = |y - x|.$$

So $\mathbf{L}_q$ is an isometry of four-space. The same argument applies to $\mathbf{R}_q$.

Now what we want is an isometry of R^3, which can be thought of as an isometry of $\mathcal{H}$ that takes R^3 to itself. Define

$$\mathbf{U}_q = \mathbf{L}_q \mathbf{R}_{q^*}.$$

Since this is a composition of isometries, it is also an isometry. (Note that since q is a unit quaternion, $q^* = q^{-1}$ is also a unit quaternion.)

Theorem 5.2.5 *For any unit quaternion q in S^3, $\mathbf{U}_q$ is an isometry of R^3. If $q = \cos\frac{1}{2}\phi + \sin\frac{1}{2}\phi V$, then $\mathbf{U}_q$ rotates R^3 about the line determined by V through the angle ϕ.*

To prove this theorem, we should check that $\mathbf{U}_q$ takes vectors to vectors. Then we should check that it does not move the vector V. Finally, we should see that it rotates vectors about the axis by the right amount. This last step is not hard if you are familiar with cross products, but it is a bit tedious without them; we'll skip the proof.

Proof

If x is a vector, we want to see that $\mathbf{U}_q(x) = qxq^*$ is also a vector. By Proposition 5.2.4, we just have to square it:

$$\begin{aligned}(qxq^*)^2 &= (qxq^*)(qxq^*) = (qx)(q^*q)(xq^*) \\ &= (qx)(xq^*) = q(x^2)q^* = (x^2)qq^* = x^2.\end{aligned}$$

The next-to-last equality comes from the fact that x^2 is real, so the commutative law of multiplication can be used. So $\mathbf{U}_q(x)$ has the same square as q and by the proposition must also be a vector.

Since $q = a + sV$ for real numbers a and s, $Vq = Va + VsV = aV + sVV = (a + sV)V = qV$. (We say that V and q *commute with each other*.) It follows that $\mathbf{U}_q(V) = qvq^* = Vqq^* = V$. So V does not move (and neither does any real multiple of it). ■

Although we have skipped a detail in the proof, it is important to say something about the angle ϕ. Notice, for example, that $i = \cos 90° + \sin 90° i$, which means that $\phi = 180°$. We can compute $\mathbf{U}_i$ easily: $\mathbf{U}_i(x) = ixi^* = ix(-i) = -ixi$. So $\mathbf{U}_i(j) = -iji - ki = -j$. Sure enough, $\mathbf{U}_i$ spins vectors $180°$ around the line through i. This doesn't match up with the role played by i as a complex number. Remember that complex multiplication by i resulted in a rotation by only $90°$ in the complex plane.

This factor of two caused confusion for many years following Hamilton's development of quaternions. Hamilton assumed that as one traveled around the unit circle in the plane containing 1 and i, the corresponding rotations would spin the sphere once around its axis, just as traveling around the unit circle in the complex plane corresponds to spinning the plane once around the origin. But in fact, the rotations $\mathbf{U}_q$ with q on the circle through 1 and i correspond to spinning the sphere twice around its axis.

More generally, the rotations determined by q and $-q$ are always the same. If we think of q and $-q$ as antipodal points on the three-dimensional sphere S^3, then this can be reformulated as the statement, the correspondence $q \longmapsto \mathbf{U}_q$ is not a 1–to–1 correspondence but a 2–to–1 correspondence between points on the 3-sphere and rotations of S. If we define the *three-dimensional projective space* $\mathcal{P}^3$, by analogy with the projective plane, as consisting of pairs of antipodal points in the three-sphere, then the correspondence becomes 1–to–1. The result of this is a description of the group of rotations of the sphere:

Theorem 5.2.6 *The group of rotations is equal to the projective space* $\mathcal{P}^3$. *The group of unit quaternions is in 2–to–1 correspondence with the rotation group.*

This theorem has very deep physical significance. If we rotate an object continuously through 360°, then every point on the object returns to its initial position. If we picture the process as tracing out a continuous path in the group of rotations, then the path forms a closed curve, beginning and ending at the identity element of the group. But in the group of unit quaternions, the corresponding path does not close up. If we go twice around the path in the rotation group, the resulting path in the quaternion group does close up. Somehow, we are distinguishing between spinning through 360° and spinning through 720°.

The twentieth century physicist Paul Dirac designed a device to illustrate this phenomenon. It consisted of a cube attached to a frame by strings running from the eight corners of the cube to the corresponding corners of the frame. If the cube is rotated through 360°, the strings become tangled up and cannot be disentangled. If the cube is rotated again by 360°, the strings seem more entangled, but in fact they can be disentangled without twisting the cube. Dirac's interest in this device relates to the observation that rotating an electron through 360° multiplies its wave function by -1.

You can perform your own experiment to demonstrate the same phenomenon Dirac studied. Take an ordinary plate, preferably with a design on it. Hold it face up in the palm of one hand. By rotating your arm, rotate the plate through 360°, keeping the design always upward. You will be uncomfortable at this point. Now rotate the plate *in the same direction* through another 360°, always keeping the plate face up. You will have to twist your wrist and arm during this process. But at the end, your arm is magically restored to its initial position. Try it!

For more information on rotations, see Simon Altmann's book [1]. (This is a rather advanced book, but the introduction is quite interesting.) An article describing *Tangloids*, a game based on the string tangling idea of Dirac, and related ideas about groups, can be found in chapter 2 of [16]. Quaternions are briefly discussed in lecture 29 of [11].

5.3 Curvature of Polyhedra and the Gauss-Bonnet Theorem

As we said in Section 4.2, Euler's theorem, $V-E+F=2$, was at one time attributed to Descartes. We will now examine the theorem Descartes did prove, which is closely related to Euler's theorem, and we will then begin to relate it to a famous theorem of Gauss, one that is at the heart of the subject known as differential geometry.

In order to form a closed curve out of a straight piece of wire, it is necessary to bend it. If we bend a wire into a polygonal curve, then the bending takes place only at the corners. The amount of bending is best measured by the *exterior angles* rather than the interior angles, since a small exterior angle corresponds to a small change in direction. If we let α_1, α_2, ..., α_n denote the exterior angles of the vertices of a convex polygon, then it is not hard to see that

$$\alpha_1 + \alpha_2 + \ldots + \alpha_n = 360^\circ. \tag{5.3.1}$$

Here is a geometric argument for this formula. For each point x on the unit circle, take the line tangent to that point. at the point of tangency, draw an arrow perpendicular to the line and pointing away from the center of the circle. We will call the line with the arrow a *contact element* of the circle.

If Σ is a convex closed curve, then we can move the contact element at x parallel to itself until it touches the curve in such a way that the entire curve lies on one side of the line and the arrow points to the other side. This gives a *support line* $\phi(x)$ of the curve.

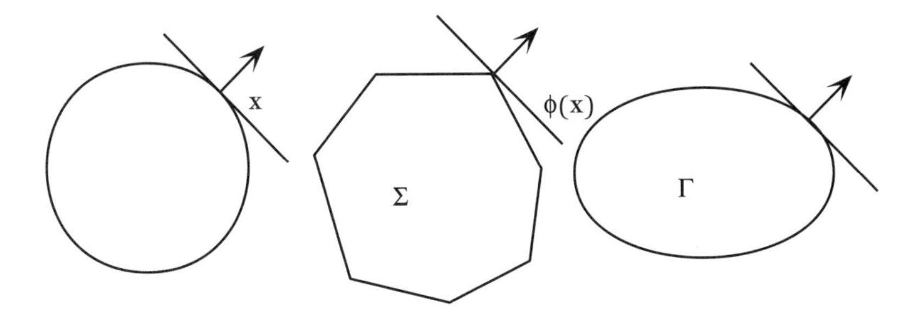

If Γ is *any* convex closed curve, we can make the same definition. Notice that in the case of a polygon, the support line usually touches the curve only at a corner; more than one support line can touch at the same point. There are support lines that touch along a whole side of Γ.

Now take all the support lines that touch at one vertex V of Σ. Two of them will be tangent to the two sides adjacent to V. The remaining support lines, which touch Σ only at V, correspond to an arc of the circle with angle θ exactly equal to the exterior angle at V.

In this way, we can divide the circle up into arcs, one corresponding to each vertex of Σ. Each arc has an angle equal to one of the exterior angles, so the sum of the exterior angles is equal to the total angle of a circle, which is 360°.

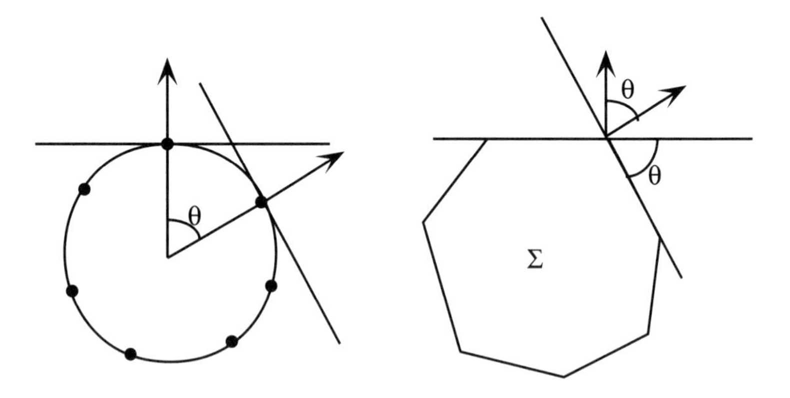

For what we will be doing in this section, it is better if we measure angles in *radians* instead of degrees. The measure of an angle θ in radians is given by measuring the length of the corresponding arc of the unit circle. Since the circumference of the unit circle is 2π, it follows that the total angle of a circle, corresponding to 360°, is equal to 2π. To convert angles from degrees to radians, just multiply the angle by $\frac{2\pi}{360}$; for instance, a right angle is $90 \times \frac{2\pi}{360} = \frac{\pi}{2}$. Using radians to measure angles, the argument above says:

Proposition 5.3.1
The sum of the exterior angles in a convex polygon, measured in radians, is always equal to 2π.

If we consider polygons that are not convex, there is a modification of Proposition 5.3.1 that is still true. When an interior angle θ at a vertex of a polygon is greater than π (180 degrees), define the exterior angle to be the *negative* number $\pi - \theta$. With this definition, we have the more general fact (not proved here):

Proposition 5.3.2
The sum of the exterior angles in a simple closed polygon is always equal to 2π.

Why is this theorem true? Here is a plausibility argument. Start on one side of the polygon and take an arrow pointing outward perpendicular to the side. As we walk around the curve, we rotate the arrow so that it always is perpendicular to each side. In the convex case, the arrow always rotates in the same direction. At a vertex where the polygon is not convex, the arrow rotates in the opposite direction. At each vertex, the amount the arrow rotates is given by the exterior angle at that vertex; negative exterior angles correspond to rotating in the opposite direction. When we return to the first side, the arrow has returned to its original position. Therefore, it has rotated through a full 2π.

Well, we don't actually know that. Maybe it rotated *twice* around the circle, for example. In fact, if we don't assume that the polygon has no self-intersections, that is exactly what may happen. In general, all we can say is that the sum of the exterior angles must be $2\pi n$, where n is an integer called the *rotation index*. We also should specify in which direction we travel around the polygon. If we go in the opposite direction, the arrow will rotate around in the opposite direction. If we adopt the standard convention that going counterclockwise around the circle increases the angle, then it is possible to prove that going counterclockwise around a convex curve always makes the arrow rotate once around in the positive direction.

This argument is not so easy to make for a nonconvex curve. In fact, once we allow for negative exterior angles, it is not so easy to decide which angles are positive and which negative, especially for curves that have self-intersections. The curve pictured below, for example, is ambiguous. Which of its exterior angles are positive and which are negative?

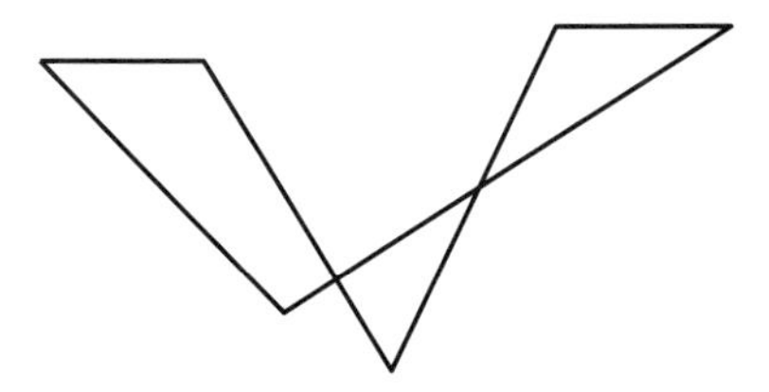

To resolve the ambiguity, we should specify which way we will travel around the polygon; this is called *orienting* the polygon. In the case of a simple closed curve, there is a standard way of doing it. By the Jordan curve theorem, the curve determines two regions in the plane, one inside and one outside. The positive orientation of the polygon is chosen by traveling around in the direction where the outward arrow perpendicular to a side is rotated counterclockwise to point in the direction we are traveling. The curve pictured above does not have an inside and an outside, so there is no clear choice of direction. The result is that we can change the sign of the rotation index of a curve by reversing its direction. So the rotation index is really defined only for oriented curves.

Problem

What is the rotation index for each of the following polygons?

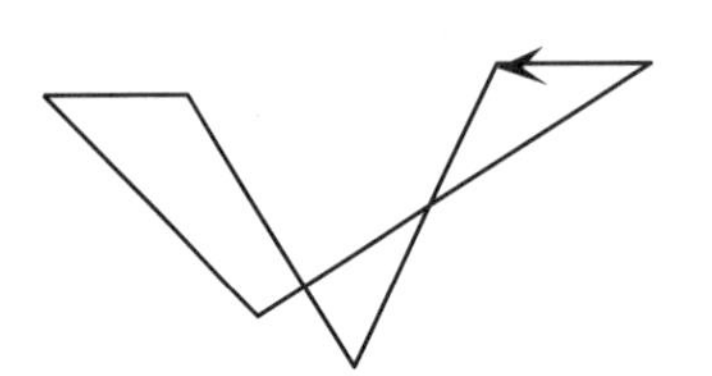

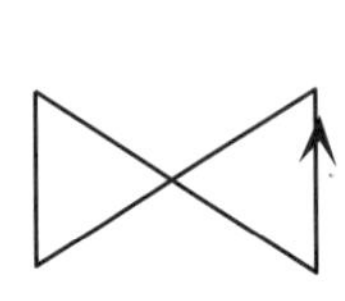

It is easy to see that the sum of the interior angles in a polygon plus the sum of the exterior angles in the polygon must be πV, where V is the number of vertices of the polygon. So for instance, a triangle must have interior angles plus exterior angles add up to 3π. This says that the sum of the interior angles of a triangle must be π (which is 180°). So we must have been relying on the fact that we are doing geometry in the Euclidean plane. What is the right analogy of this theorem in the non-Euclidean versions of the plane?

Let's look first in the hyperbolic plane. There, the sum of the angles in a triangle is always less than π. In fact, we saw in Section 2.2, that π minus the sum of the angles in a triangle (the *defect* of the triangle, now translated into radians) is proportional to the area of the triangle. Again, the sum of the interior angles in a triangle plus the sum of the exterior angles must always be 3π; the Euclidean argument hasn't changed. Therefore, we have the following principle:

Lemma 5.3.3

If $\triangle ABC$ is a triangle in the hyperbolic plane, then the sum of its exterior angles is always greater than 2π by an amount proportional to the area of the triangle.

Proof

(Sum of exterior angles in $\triangle ABC$) $= 3\pi -$ (the sum of the interior angles) $= 3\pi - (\pi -$ the defect of the triangle$) = 2\pi + C\times$(area of the triangle), where C is the constant of proportionality. ■

You may be wondering about the mysterious constant C in this lemma. If we change the unit of length (and therefore the unit of area) in the plane, we will change the value of the area but not the angles. Consequently, we will be changing the value of C. In the "standard" model of the hyperbolic plane, we choose the unit so that $C = 1$. Note that in that case, any triangle will have area no larger than π.

Problem

Let's assume for convenience that $C = 1$. Show that the sum of the exterior angles of any polygon is always equal to 2π + the area of the polygon. [HINT: Cut the polygon up into triangles.]

There is an exact analogue of the lemma above that applies to polygons in the sphere. Now, of course, the sum of the interior angles in a triangle is larger than π, so the sum of the exterior angles in a triangle will be smaller than 2π. The following theorem is a basic fact in spherical geometry:

Theorem 5.3.4 *The sum of the exterior angles of any polygon in the unit sphere plus the area of the polygon is always equal to 2π*

Because we chose the sphere of radius 1, the constant of proportionality in the theorem above turns out to be 1. If we scaled the sphere, then the area would also scale, and so, as in the hyperbolic case, we would need to put in a new constant of proportionality. The scale factor turns out to be the reciprocal of the square of the radius. In other words, the sum of the exterior angles plus $\frac{1}{R^2} \times$ the area adds up to 2π. As R gets larger, this constant shrinks more and more. We can imagine a plane as being a "sphere of infinite radius," in which case the term $\frac{1}{R^2}$ could be thought of as 0. Then the sum of the exterior angles would be exactly 2π. The quantity $\frac{1}{R^2}$ is the *curvature* of the sphere of radius R. The plane has curvature 0. By analogy, the standard model of hyperbolic space must have curvature -1; we may fancifully think of it as a sphere of imaginary radius i, whatever that means! The hyperbolic plane is sometimes called a "pseudosphere."

Now let's move away from the problem of polygons and think instead about polyhedra. What is the analogous theorem, if any, for convex polyhedra? To start, define the *contact element* at the point x of the unit sphere S to be the plane tangent to the sphere together with an arrow perpendicular to the plane pointing away from the sphere. If Σ is a convex polyhedron, a *support plane* at a point p is a plane that passes through p for which Σ lies entirely on one side of the plane. Attach to this plane an arrow at p that points perpendicular to the plane and away from Σ. Now given any point x of the unit sphere, we can translate the contact element at x parallel to itself until it becomes a support plane at a point $p = \phi(x)$ with the arrow poining outward.

As in the case of polygons, the support plane will usually touch only at a vertex of the polyhedron. To take a concrete example, consider a cube, resting flat on a table. The north and south poles of the sphere correspond to the top and bottom support planes of the cube. There are four equally spaced points on the equator corresponding to support planes tangent to the four lateral faces of the cube.

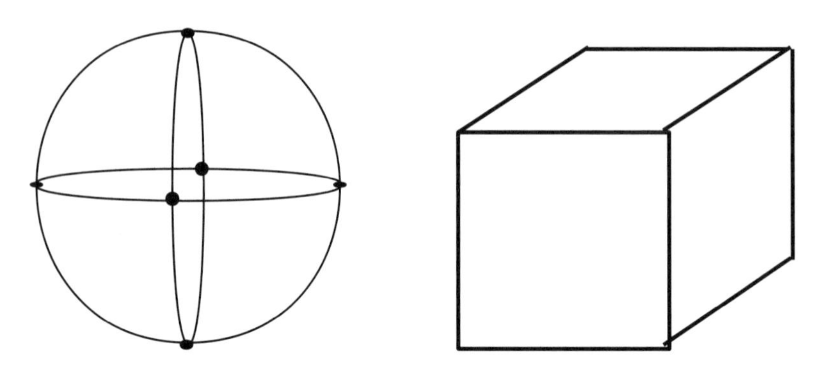

All the remaining points on the equator correspond to support planes that touch the cube along one of the vertical edges. Each great circle arc connecting one of the four points on the equator to one of the poles corresponds to support planes that touch the cube along one of the horizontal edges. This divides the sphere into eight congruent triangles, each of which corresponds to the support planes that touch one of the eight corners.

How exactly does this correspondence go? Let's trace the triangle corresponding to one corner V of the cube. Take a vector (an arrow) perpendicular to one face of the cube adjacent to V and slide it over to an edge emanating from V. Rotate it perpendicular to the edge until it becomes perpendicular to the adjacent face. Now slide it along the face to the next edge from V. Repeat the process until we get all the way around the vertex. Each great circle arc on the sphere is traced out by moving a vector perpendicular to one edge. Its length is given by the *dihedral angle*, that is, the angle between adjacent faces. The interior angle α of the triangle on the sphere between this great circle arc and the next one is determined by how much we change direction in moving from one edge of the cube to the next. This angle is just the supplementary angle to the angle a of the face at the vertex V. In the case of the cube, these are both right angles. Thus, the angles of the polygons at the vertex V are equal to the exterior angles of the spherical triangle corresponding to V.

By Theorem 5.3.4, the area of each spherical triangle is 2π minus the sum of the exterior angles. The observation above says therefore that the area of the triangle is 2π minus the sum of the angles at V. We define the *defect* at a vertex V of a polyhedron to be this quantity 2π − sum of the angles at V.

Each of these eight spherical triangles has area equal to one-eighth the total area of the sphere, or $\frac{1}{8} \times 4\pi = \frac{\pi}{2}$. Correspondingly, the defect

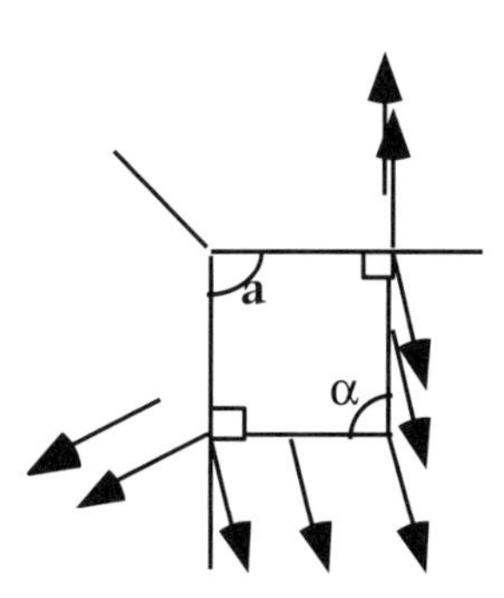

at each of the eight vertices of the cube is equal to $\frac{\pi}{2}$. This is an instance of Descartes's Theorem, which we are now ready to state.

Theorem 5.3.5 (Descartes's Theorem) *If Σ is any convex polyhedron, then the sum of the defects at all of the vertices of Σ is 4π.*

We have all we need for a proof of this theorem. If Σ is a convex polyhedron, then we divide up the unit sphere into polygons, one for each vertex, where each polygon corresponds to the set of support planes to a vertex of Σ. The area of each polygon is equal to the defect of the corresponding vertices. The sum of the areas is the area of the sphere, which is 4π.

However, there is another proof of this theorem that does not rely on this division of the sphere. Instead, we will see that the theorem is a direct consequence of Euler's theorem. In fact, this is a two-way street: if we assume Descartes's theorem, we can prove that for any convex polyhedron $V - E + F = 2$. As Malkevich points out in his article ([31], p. 86), this has led to the erroneous impression that Descartes could easily have discovered Euler's formula. But that would have required Descartes to think of a polyhedron as a *combinatorial* object, rather than a geometric object, a major intellectual leap at the time.

Take a convex polyhedron Σ. To simplify things a bit, we can divide each face up into triangles by drawing diagonals. For example, a cube can be thought of as having twelve triangular faces, each an isosceles right triangle. Let F be the number of (triangular) faces of Σ. Then the number of sides of polygons is $3F$, and so the number of edges is given by $E = \frac{3}{2}F$. If V is the number of vertices, then Euler's theorem says that $2V - 2E + 2F = 2V - F = 4$.

The sum X of the defects at the vertices is given by summing the quantities (2π − angles at V). This gives $2\pi V$ − (the sum of all the angles at all the vertices). This sum can be computed in a different way, however. Instead of adding up the angles, vertex by vertex, we add up the angles triangle by triangle! The sum we get is exactly πF. So

$$X = 2\pi V - \pi F = 4\pi.$$

To close this section, we will look at an analogue of this theorem that holds for convex surfaces that are *smooth*, which means (roughly speaking) that there are no corners, but instead a plane tangent to the surface at each point. Examples of such objects are spheres, ellipsoids (surfaces given by an equation of the form $\frac{x^2}{a^2} + \frac{y^2}{b^2} + \frac{z^2}{c^2} - 1 = 0$), and more generally, the set of solutions to some equation $F(x, y, z) = 0$, where F is a "nice" function and where the solution set happens to be a convex surface.

The function ϕ that assigns to each point x in the unit sphere the point $\phi(x)$ on Σ that has a support plane perpendicular to the line from the origin to x is well-defined. If P is a set of points on the sphere with area $A(P)$, then we say that the *integral curvature* of $\phi(P)$ is the quantity $A(P)$. The integral curvature of a piece of Σ measures how much the piece bends around. With this definition, we can say that the total integral curvature of a convex surface is always equal to 4π. This is a special case of what is known as the *Gauss–Bonnet theorem*.

Of course, at the moment this seems like a self-evident fact and therefore not important. To appreciate the result, let us look back at Descartes's theorem. With the same definition of integral curvature we can see that if we take a subset P of a polyhedron Σ that has no vertices, then its integral curvature is 0. This is because ϕ takes all of the unit sphere to the vertex set except for some points and circle arcs, and those don't have any area. In fact, if $\kappa(P)$ is the integral curvature of a set P, then

$$\kappa(P) = \kappa(V_1) + \kappa(V_2) + \ldots + \kappa(V_n),$$

where $V_1, V_2, \ldots V_n$ are the vertices contained in P. For a vertex V, $\kappa(V)$ is the defect of the vertex V.

Now, the defect of a vertex can be measured by looking only at the geometry of the polygons that share the vertex. If we imagine a "two-dimensional" being living on the surface Σ equipped with measuring instruments, we can imagine that this being could detect the defect at a vertex by making measurements. To put this in more technical terms, we say that the curvature at a vertex is *intrinsic* to the geometry of a surface. It does not depend on how the polyhedron sits in space.

For example, take four triangles meeting at a vertex V. The support planes to this pyramid at V determine a spherical quadrilateral whose area is the defect at V. If we pinch the pyramid, so that the dihedral angles change but the triangles themselves do not change, then the spherical quadrilateral will change. Yet its area stays the same!

By contrast, dihedral angles are not intrinsic to a polyhedral surface. For if we take two polygons that are connected along an edge, we can change the dihedral angle between them without stretching or tearing the polygons.

Gauss, who studied the geometry of smooth surfaces in the early nineteenth century, proved a beautiful theorem that he called the *Theorema Egregium*, or Excellent Theorem. (Nowadays the word egregious has taken on a negative meaning!) His theorem states that the curvature of a smooth surface is intrinsic, so that it can be determined from the internal geometric relations of points on the surface. The Gauss–Bonnet theorem says that if we change the intrinsic geometry of a surface (by stretching and distorting it), we can change the curvature in different places but the total integral curvature cannot change.

The Gauss-Bonnet theorem says more than that. Even if the surface is not convex, the formula continues to be valid. This is reminiscent of the principle for closed curves in the plane, but there is a difference. If the surface has no self-intersections, it may still have total curvature different from 4π. In fact, the total curvature of a surface must be $2\pi\chi$, where χ is the number Euler's theorem predicts for $V - E + F$. For example, the *torus* is the surface of an inner tube or a doughnut. The Gauss- Bonnet theorem predicts that the total integral curvature of such a surface must be 0. Note that as in the case of curves, it is necessary to talk about negative curvature for this formula to make sense. If a surface has only positive curvature, then it turns out to be convex.

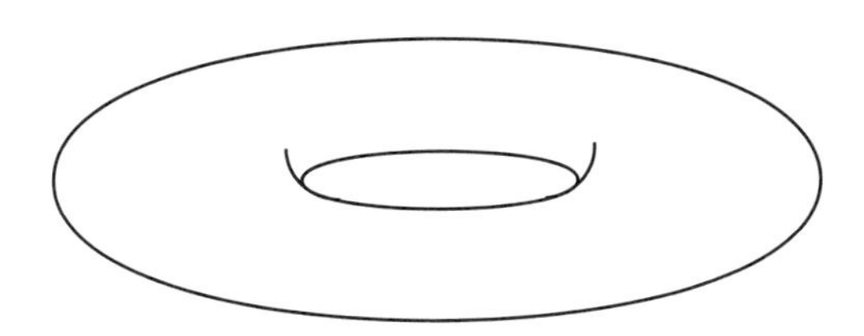

Problem

Divide the surface of a torus (pictured above) into regions (with curved sides) in such a way that any two regions that touch, touch along a whole side or at one point. Compute V, E, F for your diagram, and determine the value of $V - E + F$.

Problem

One way of doing the previous problem is to imagine a torus as being constructed from a sheet of rubber in the form of a rectangle by gluing opposite pairs of sides together. Divide a rectangle into polygonal regions, cutting up the edges of the rectangle into segments and remembering to match corresponding segments. What is the smallest number of such regions into which you can divide a rectangle so that no two regions share more than one edge? How many colors do you need to color the resulting diagram so that regions with a common border have different colors?

6 CHAPTER Geometry of Space

6.1 A Hint of Riemannian Geometry

In the last section of Chapter 5, we referred to curvature as being *intrinsic*. We will explore the notion of intrinsic and extrinsic properties of a geometric object in this section, ending with a rough description of Bernhard Riemann's reformulation of geometry.

Let's begin with a puzzle, quoted from *Mathematical Puzzles* by Geoffrey Mott-Smith (Dover, 1954), called "The Spider and the Fly" (p. 60).

A spider lived in a rectangular room, 30 feet long by 12 feet wide and 12 feet high. One day the spider perceived a fly in the room. The spider at that time was on one of the end walls, one foot below the ceiling and midway between the two side walls. The fly was on the opposite end wall and one foot above the floor. The spider cleverly ran by the shortest possible course to the fly, who, paralyzed by fright, suffered himself to be devoured. The puzzle is: What course did the spider take and how far did he travel? It is understood he must adhere to the walls, etc.; he may not drop through space.

Problem

Before reading on, solve the puzzle!

The fly, being capable in calmer moments of flying at will through space, views the walls of the room as part of a three-dimensional world.

By contrast, the spider, constrained by nature to cling to the walls of the room (except when it spins a web!), has a different view of it. To the spider, the walls form a two-dimensional world made up of six rectangles connected along edges. If we were to unfold some of the walls and lay them flat on a plane, the spider's path along the walls could be traced out in this plane; its length would not change.

Of course, it is necessary to sever the connections between some of the rectangular sides of the room to lay them out, but as long as we don't break the connections that the spider will use in its walk, it makes no difference. If, for example, the spider crawls along the ceiling to get to the fly, then the path it will take will look like a straight line A when we unfold the walls as illustrated below. It is then obvious that the distance the spider crawls will be 42 feet.

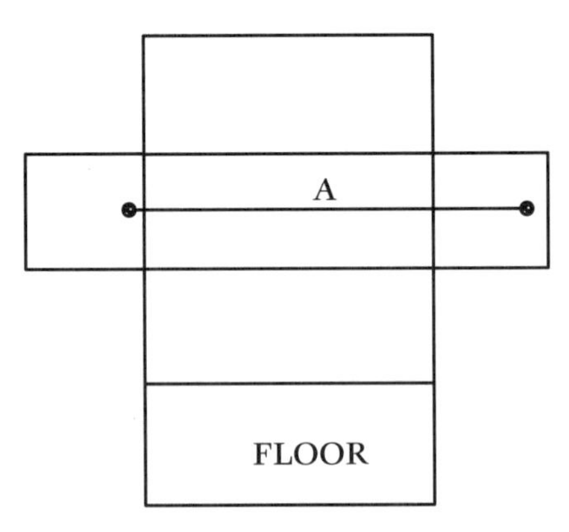

The spider could also crawl along the ceiling and one of the long walls on its way to the fly. This path is the straight line B illustrated below, whose length turns out to be approximately 40.7 feet. If the spider chooses path C below, which uses the floor as well, the length of the path is exactly 40 feet. This is the actual shortest path from spider to fly.

How do we know that the spider will crawl on a straight–line path and not a zigzag? More generally, if instead of a rectangular room, we had a spider and a fly on some polyhedron, how do we know that the shortest route joining them would look like a straight line if we laid out the faces in order? Before examining this problem, let us first define the polyhedra for which this question makes sense. A *(Euclidean) polyhedral surface* is a finite collection of polygons glued together along edges according to the following rules:

1. Each polygon is congruent to a convex polygon in the Euclidean plane.
2. Any two polygons that have a point in common meet either at a common vertex or along a common edge.

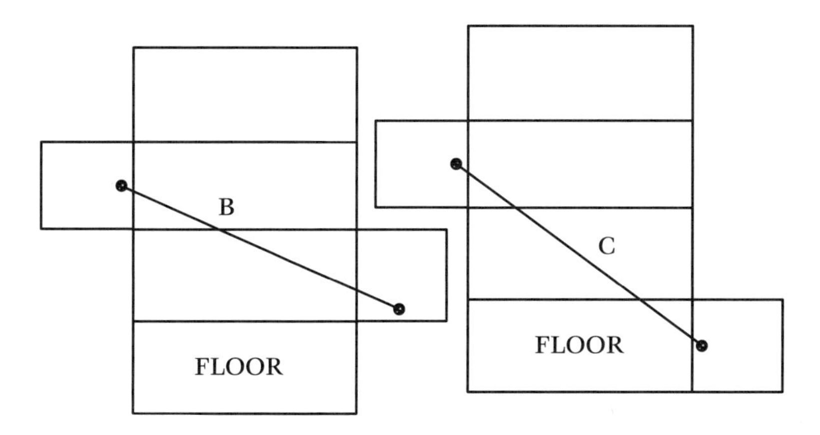

3. If two polygons have an edge in common, then the edge has the same length in each polygon.
4. Every edge of a polygon is an edge of exactly one other polygon.
5. If $\alpha_1, \alpha_2, \ldots, \alpha_n$ are all the polygons that share a vertex V, then we can renumber the polygons in such a way that α_i and α_{i+1} share an edge for $1 \leq i < n$ and α_1 and α_n share an edge.

A convex polyhedral surface is an example of a Euclidean polyhedral surface. The fifth rule prevents us from encountering such objects as two tetrahedral surfaces glued together at one vertex. If the first rule is replaced by the statement that each polygon is congruent to a convex polygon in the hyperbolic plane (respectively, the sphere), then the result is a *hyperbolic* (respectively, *spherical*) polyhedral surface. Note that it is not necessary to assume that the polyhedral surface can actually be assembled in space the way the rectangular walls of the room can be assembled to form the room. Consequently, such a surface is sometimes called an *abstract* polyhedral surface. Since we may not have a model of the polyhedron sitting in space, the *only* properties we can study are intrinsic ones.

Suppose Σ is a (Euclidean) polyhedral surface. What does the shortest path joining two points A (= spider) and B (= fly) look like? In order to answer this, we need to assume that a shortest path exists. This is by no means obvious; it can be proved, but the proof requires some difficult mathematical arguments. (You may find them in a text on differential geometry.) But now that we have made the assumption, suppose γ is a shortest path from spider to fly. Pick two points X and Y along the path close to each other. If they are in the same polygon, then the part of γ joining them must be a straight line segment, since otherwise we could replace that portion of the path by a shorter one. (Here we are using the

Euclidean fact that the shortest path between two points in the plane is a straight line segment.

If X and Y are in different polygons that share a common edge, and the path γ crosses the edge at some point Z, put these two polygons down adjacent to each other in the plane. If the straight line joining X to Y stays in the two polygons, then again it is shorter than the piece of γ. If not, then at least the line segment from X to Z and the line segment from Z to Y form the shortest path from X to Y going through Z. But then there is a shorter path from X to Y that goes through one of the endpoints of the common edge.

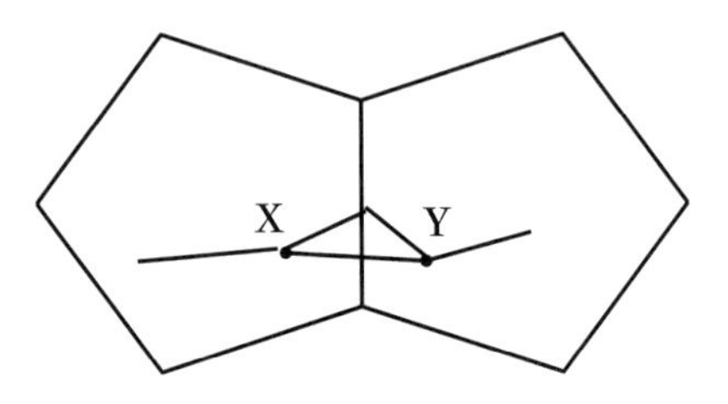

(Why is this true? Intuitively, connect the points with a piece of a very elastic band and let the band shrink. If the band were just allowed to move anywhere in the plane, it would snap to a straight line. If the band is forced to pass through the line segment between the two polygons, it will pull off to one end or the other.)

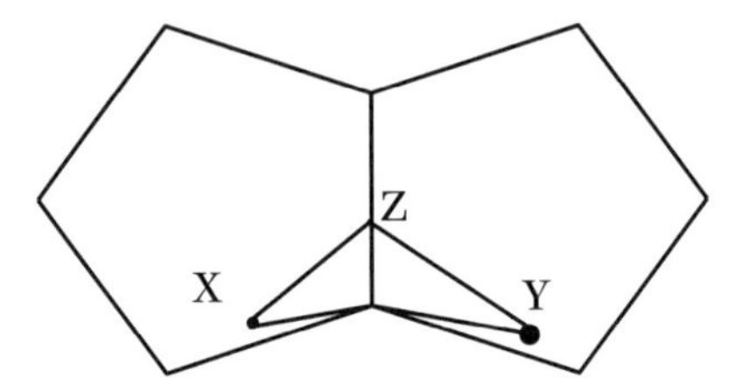

What we have decided so far is that if γ really is a shortest path, then the part of it inside any one polygon will be a straight line, and if it crosses a border anywhere except at a vertex, the crossing will be such that the two segments on opposite sides of the edge fit together to form a straight line segment. There is, however, the possibility that a shortest path will go through a corner. Let k be the angle deficit at the corner, and assume that $k > 0$. Then we can put all of the polygons that meet at this corner down onto the plane so that they share the corner. For instance, if we are on a dodecahedron, we can put down the three pentagons so that they do not overlap.

In general, there will be more than one way to lay out the polygons that meet at a vertex onto the plane. In our example, there are three different ways the three pentagons can be placed, corresponding to the three edges along which we would have to cut in order to flatten the corner out. Now, it is always true that at least one of the possible ways will have the property that the line segment joining X to Y in the plane stays within the polygons. (This is left as a problem for the reader to verify.) But that means that there is a shorter path from X to Y that does not go through the vertex.

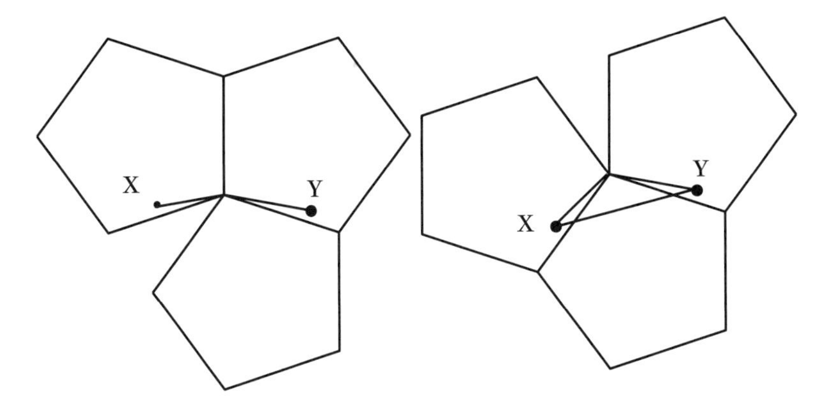

To summarize, we have the following theorem about shortest paths:

Theorem 6.1.1 *A shortest path on a Euclidean polyhedral surface joining two points X and Y is made up of straight line segments that meet edges of the polyhedron in such a way that adjacent segments either fit together to form a straight line when the adjacent polygons are laid out on the plane or meet at a vertex with negative angle deficit.*

Problem
Justify the argument above that says that if a path goes through a vertex with positive deficit it is always possible to find a shorter path that misses the vertex.

Problem
Suppose six right triangles meet at a vertex, as pictured below, in some polyhedron. Let X and Y be points in two opposite triangles (separated by two triangles in either direction going around the vertex). Show that if the shortest path stays within the triangles, then it must go through the vertex at the center. Construct an example of a polyhedron with such a vertex, in which the shortest path between the points X and Y does not *go through the vertex.*

In the picture above, it is not hard to see that if X and Y were picked very close to the central vertex, then the shortest path would of necessity go through that vertex. That is because any path which does not go

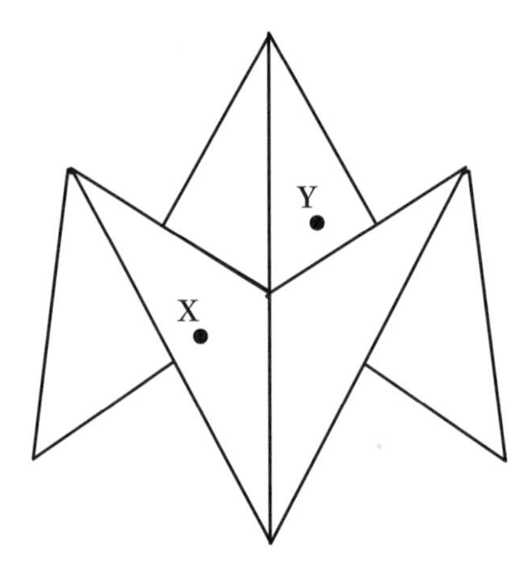

through the vertex would have to leave the triangles, and just getting out to the border would be too big a detour. We can restate this observation in another way. A two-dimensional being living on the surface of a polyhedron will be able to detect a nearby vertex with positive defect by the fact that shortest paths tend to avoid going through it. The being will be able to detect a nearby vertex with negative defect by noticing that shortest paths tend to go through it. But the being will not be able to detect an edge. The defects at vertices are part of the *intrinsic geometry* of the polyhedron, while the edges are part of the *extrinsic geometry*.

If a vertex has defect exactly 0, then its geometry will be Euclidean. For example, if we divide a polygon into smaller polygons by adding vertices and edges, then the new vertices will not be detectible by intrinsic measurements. Is it possible to have a polyhedral surface with 0 defect at *every* vertex? Yes it is. Here is an example: Take a rectangle and divide it into nine similar, smaller rectangles. Now glue the top edge of the rectangle to the bottom edge and glue the left edge to the right edge. The resulting polyhedral surface has nine rectangles with two rectangles meeting at each edge. The angle sums at each vertex are 2π. For example, the rectangles numbered 1, 3, 7, and 9 come together around one vertex.

A two-dimensional being living on this surface would not be able to detect any deviation from the rules of Euclidean geometry as long as the being did not try to travel long distances. Notice, however, that a straight path drawn horizontally along the polyhedron closes up to form a closed curve, (a *closed geodesic*), as does a vertical straight path. This is reminiscent of the phenomenon of great circles on the sphere, but the similarity ends here. Diagonal lines behave quite differently from great circles on the sphere. For example, there is a straight path from the point X through the point Y, then through Z, then returning to X. It crosses

9	7		
3	1	2	3
	4	5	6
	7	8	9

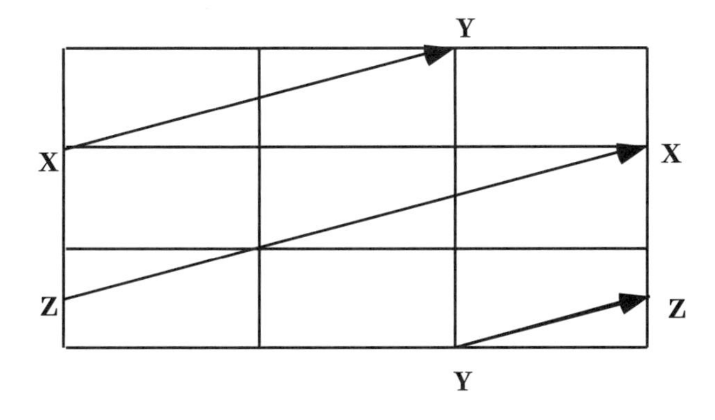

each horizontal closed geodesic once, but it crosses each vertical closed geodesic *twice*!

If we attempt to build a model of this surface in space, we can take a rectangular piece of paper and glue the top edge to the bottom, forming a tube. Now we need to attach the left end of the tube to the right end. This can only be done by crumpling the tube; nevertheless, we can connect them in our imagination. What we get is a *torus*. From the last problem of Chapter 5, or from the construction above, we know that for such a surface $V - E + F = 0$. This could also be predicted from the abstract version of Descartes's theorem:

Theorem 6.1.2 (Polyhedral Gauss–Bonnet Theorem) *Let $\mathcal{X}$ denote the value of $V - E + F$ for a polyhedral surface. Then the sum of the defects at the vertices of any Euclidean polyhedral surface is $2\pi\mathcal{X}$.*

The quantity $\mathcal{X}$ is called the *Euler characteristic* of the surface. The Euler characteristic of the sphere is 2, while that of the torus is 0. The proof of the theorem above is identical with the second proof we gave of Descartes's theorem in Chapter 5.

Are there other examples of Euclidean polyhedral surfaces that have 0 defect at every vertex? In fact, there are infinitely many. Replace the rectangle above by any other rectangle, or more generally by a parallelogram. The result will be a torus. Is it the same torus or are they different? If the parallelogram has a different area, then we can tell the two tori apart. By scaling all the dimensions, we can get a similar surface with any area we want. Let us assume that the area is always fixed at 1. Then could a being living on one of the two surfaces tell the difference between them by some measurement on the surface?

Again the answer is yes. Our rectangular torus gives us two families of closed geodesics that are everywhere perpendicular to each other. If we used a different rectangle, the lengths of these closed geodesics would be different. In particular, the vertical geodesic in our example above is the shortest closed geodesic. Its length is an intrinsic geometric quantity.

If we took a parallelogram instead of a rectangle, then in general if we take the shortest closed geodesic, there will be a family of such curves parallel to it. But the curves perpendicular to them will *not* usually close up after going once around. In the example pictured below, the shortest closed geodesics are horizontal; a single vertical geodesic crosses each of them four times before closing up.

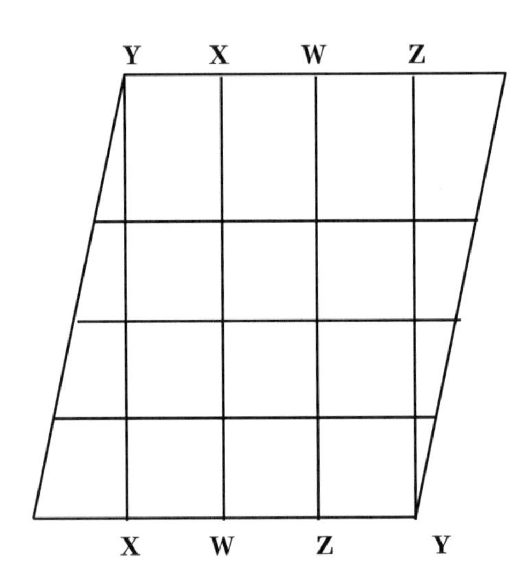

Problem

Divide a regular hexagon up into polygons. Now glue opposite pairs of edges together to form a Euclidean polyhedral surface. Study the properties of this surface.

Not every Euclidean polyhedral surface with 0 defects is a torus. If we change the gluing rule for the sides of the rectangle so that the left side is given a twist before being attached to the right side, the resulting surface is called a *Klein bottle*. Like the projective plane, the Klein bottle is a cousin of the Möbius band, but it has Euler characteristic 0, whereas the projective plane has Euler characteristic 1. It turns out that the torus and the Klein bottle are the only surfaces that have Euler characteristic 0.

Problem

Investigate the possibility of tiling a Euclidean polyhedral surface with 0 *defects by regular polygons. Relate this problem to the problem of tiling the plane.*

Our discussion so far has concerned the geometry of polyhedra. Now let us shift our attention to smooth surfaces, which we discussed briefly at the end of the last chapter. We want to consider shortest paths joining two points X and Y on such a surface. Because a smooth surface in Euclidean three-space is curved, it does not usually contain any straight lines. How then do we talk about shortest paths? If γ is a curve in space, we can inscribe a polygonal path in it by choosing some points on the curve and connecting consecutive points with line segments. The length of the curve should be more than the sum of the lengths of such inscribed paths. This allows us to define the length of γ to be the *least upper bound* of the lengths of all possible inscribed paths.

This idea of length goes back at least to Archimedes, who inscribed polygons in a circle to estimate its circumference (and compute the value of π). A mathematical treatment of length of curves requires calculus, but we can proceed here on an intuitive basis, pretending that we know how to measure the length of a curve. (If we had a surface in front of us, for instance a globe, we could lay a piece of string over a curve, then straighten the string and measure it.) Incidentally, complicated curves can be infinitely long; an example is the perimeter curve of the Koch snowflake from Chapter 2.

A *geodesic* on a surface is a curve that is the shortest path between any two points on it that are close together. Why do the points have to be "close together"? A good example is the sphere, where a great circle arc is a geodesic. Nearby points on the arc have it as the shortest path, but if we go too far, there will be a shorter arc gotten by going around the great circle in the opposite direction.

We can imagine two-dimensional beings living on the surface and capable of making geometric measurements of lengths and angles. Actually, since human beings live on the surface of the earth, we do not need such a vivid imagination. When we measure the length of a stretch of a path, we make measurements based on the misleading impression we have that the earth is flat. We perceive of a geodesic as a straight line, when actually it is part of a curved line in space that lies on the surface of the globe.

By the intrinsic geometry of a surface, we mean those geometric properties that can be studied using length and angle measurements on the surface alone. So the circumference around the equator of the earth is intrinsic, but the diameter of the equator (thought of as a curve in space) is extrinsic.

Why restrict ourselves to two dimensions? Let us now imagine a three-dimensional "space" in which we can do geometry. We assume that we can measure lengths of curves in the space and also angles. Any properties of objects that only depend on such measurements will be intrinsic. What are the general principles, the postulates, of this geometry?

In his great lecture of June 10, 1854, *Über die Hypothesen, welche der Geometrie zu Grunde liegen* (On the Hypotheses which lie at the Foundations of Geometry), Georg Friedrich Bernhard Riemann developed a new view of geometry, which we now call *Riemannian Geometry*. He proposed a general view of geometry in which one could speak of three-dimensional (or in fact, any-dimensional) spaces, in which one could make measurements of lengths and angles. His guiding principle was that in a very tiny portion of space the rules of geometry should be "approximately Euclidean." What he actually proposed was that at the *infinitesimal* level, the geometry should be exactly Euclidean. In such a space, a tiny being would see geometric figures as conforming to the laws of Euclidean geometry, although on a very large scale these laws might fail.

If we look at the parallel postulate as telling us that the sum of the angles in a triangle is always exactly 180°, then this is either true or false. But consider a tiny triangle on the sphere. The sum of the angles in the triangle is greater than 180°, but by an amount proportional to its area. So if the area of the triangle is very small (compared to the total area of the sphere), then the discrepancy will be in the range of experimental error. This is our experience on Earth: a small triangle laid out by surveyors on the surface of the earth will appear to have angle sum 180°, because our instruments cannot measure the actual sum carefully enough to tell the difference. Euclidean geometry is perfectly reliable for measurements on the earth's surface, provided that we take only a very small part of the earth.

Riemannian geometry is the study of the properties of spaces that on a microscopic level are Euclidean but on the macroscopic level may be quite different.

Problem

Assume that the earth is perfectly round and has a radius of 4000 miles. If we construct an equilateral triangle on the surface of the earth whose angle sum is 181°, approximately how much area would the triangle cover? Compare this area with some geographic region.

6.2 What Is Curvature?

In this section we will briefly explore the notion of curvature for a smooth surface. To do this, it is helpful to begin with the idea of the curvature of a smooth curve in the plane. Such a curve σ has at each point a tangent line, and we may, as we did in Section 5.3 in the case of convex curves, draw an arrow perpendicular to that line. There are two choices for such an arrow; the important thing is to make the choice consistent as we travel around the curve. The arrow is called the *normal vector* to the curve.

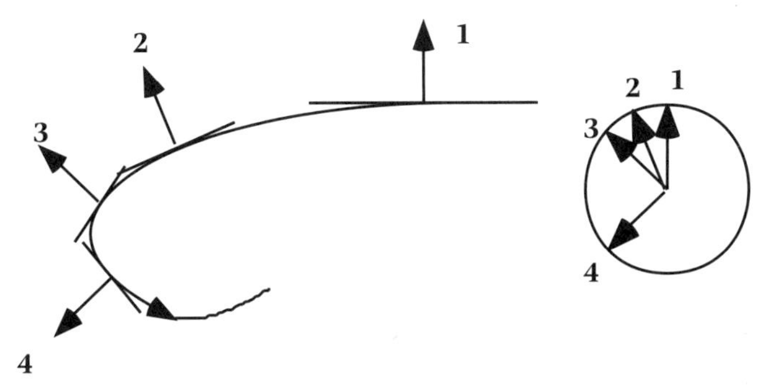

Corresponding to each point P on the curve we get a point $\gamma(P)$ on the unit circle by carrying the normal vector parallel to itself until its tail is at the center of the circle. (We assume that the vector has length one, so its head touches the circle.) This correspondence γ is called the *Gauss map* of the curve. The *geodesic curvature* $\kappa(P)$ of the curve at the point P measures the rate at which the arrow is rotating as we move along the curve through the point P. That is, if we take a small piece of the curve of length ℓ with P in the middle, and if we measure the total change ϕ in the angle at which the vector points between the two ends of the piece, then the ratio $\frac{\phi}{\ell}$ is the average curvature of that piece of curve. If we take tinier and tinier lengths, this average will more and more resemble a fixed number κ.

What we are really doing is computing a *derivative* of the Gauss map, which requires some technical assumption about the curve ("smoothness"). But since this requires calculus, we will try to leave this at an intuitive level here. In the picture above, the curvature κ is relatively small at point 1 and 2 and gets larger at 3 and 4. A circle of radius R has the same curvature at every point, namely $\kappa = \frac{1}{R}$. Larger circles have smaller curvature. A straight line has curvature 0. If we take a small piece of a curve through a point P, we can approximate it by an arc of a circle. To do this, pick a point on each side of P and very close to it. There is a circle (or possibly a straight line) that passes through these three points. As we allow these points to get closer and closer to P, we get better and better approximations to the curve. The "best-fitting" circle, called the *osculating* circle, has radius $R = \frac{1}{\kappa(P)}$. If $\kappa = 0$, the osculating circle is actually a straight line.

There is a slightly more precise definition of curvature that allows curvature to be positive or negative, depending on the choice of normal vector: If we reverse the choice of normal vector, then the sign of the curvature will change. With a circle, the inward-pointing normal determines the positive sign for the curvature. In general, if the curve lies on one side of its tangent line and the normal vector points to the same side, then the curvature is positive. The curve can cross its tangent line at a place where the curvature is zero.

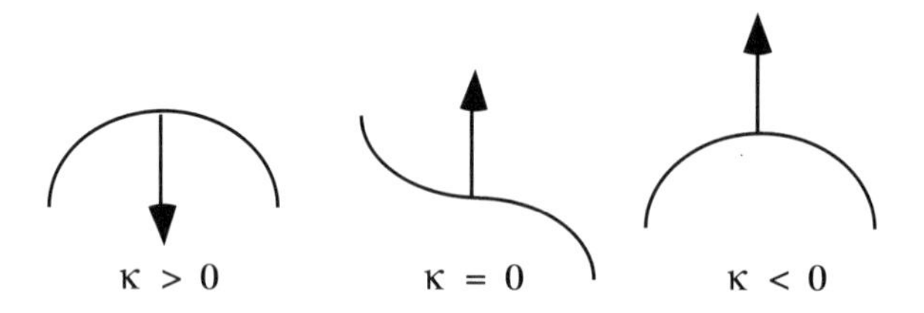

Now let's look at curvature for surfaces. Leonhard Euler dealt with the idea of curvature of a surface through the examination of *plane sections*. Suppose that we have a smooth surface Σ and a point P on the surface. The line ℓ through P perpendicular to the tangent plane at P is called the *normal line*. To describe the way the surface curves at P, we can form the intersection of the surface Σ with a plane that contains ℓ and then compute the curvature κ of the intersection curve, called a normal section, in the plane. Euler showed[1] that each of these normal sections has curvature given by the formula

$$\kappa = \kappa_1 \cos^2 \theta + \kappa_2 \sin^2 \theta,$$

where κ_1 and κ_2 are the largest and smallest curvatures, and θ is the angle between the plane corresponding to κ_1 and that for κ. In particular, the planes corresponding to κ_1 and κ_2 are orthogonal. So Euler established that in some sense the curvature of a surface at a point could be described by means of two numbers. (In fact, this statement is more precisely due to

1. *Recherches sur la courbure des surfaces*, Histoire de l'Académie Royale des Sciences de Berlin, **16** (1760), 119-143.

Meusnier, who showed that even the curvatures of *nonnormal* sections, formed by using planes through P not containing the normal line, are determined by this pair of numbers.)

There is an ambiguity in the definition of the normal curvature, coming from the fact that the sign of the curvature of a curve in the plane depends on which normal we use. If we specify a normal vector to the surface instead of just a normal line, then it can be used to fix the sign of the curvature of a section. For example, on a sphere the inward-pointing normal corresponds to the curvatures of normal sections (that is, great circles) being positive.

The modern theory of surfaces originates in the magnificent work of Carl Friedrich Gauss, who set forth many of the central ideas in his 1827 Göttingen lecture *Disquisitiones Generales Circa Superficies Curvas*. Gauss defined a function from the surface Σ to the unit sphere, exactly analogous to the function described above from curves to the unit circle. The function assigns to each point P on the surface the point $\Gamma(P)$ on the sphere whose direction corresponds to the normal direction to the surface. (We assume that a normal direction to the surface has been specified, such as the outward direction for a convex surface.)

Using this function, which is now known as the *Gauss map*, he defined the *integral curvature* of a region of Σ to be the area of its image under the map. This is the same definition we used in Chapter 5 for polyhedra. Then he defined the *measure of curvature* $K(P)$ at a point P to be the limiting value of the ratios of integral curvature to area for small regions surrounding P. This is the analogue of the geodesic curvature for a curve, with area replacing length. Then Gauss related this number to the curvatures κ_1 and κ_2 by the simple formula

$$K(P) = \kappa_1 \kappa_2$$

Notice that if we switch our choice of normal direction, then *both* κ_1 and κ_2 change signs, so the sign of K is not affected by this choice. K is usually known as the *Gaussian Curvature*. Unlike the case of curves, the sign of the Gaussian curvature does not depend on a choice.

Now we have a very pretty geometric interpretation of curvature; in particular, we can see a qualitative difference between positive and negative values of K manifested in the shape of the surface. How do we get a negative value? A convex surface will always have positive curvature, because all of the normal sections curve in the same direction. An example of a negatively curved surface is a saddle (pictured below). As we vary the choice of normal section, the curves change the sign of the curvature.

This particular saddle is the graph of the function $z = f(x, y) = x^2 - y^2$. At the origin $(0, 0, 0)$, this surface has a tangent plane that is horizontal.

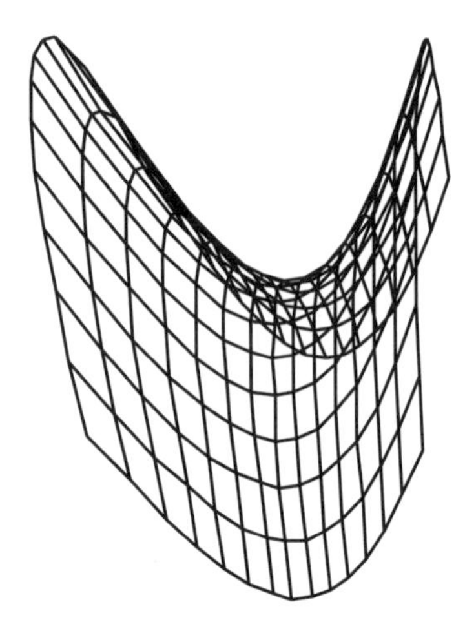

(This may surprise you, since it means that the tangent plane cuts the surface near the point of tangency.) The z-axis is the normal line to the surface. The intersection of this graph with the xz-plane is the parabola $z = x^2$. If we choose the upward-pointing normal to the surface, then this curve is positively curved. The intersection of the graph with the yz–plane is the downward-pointing parabola $z = -y^2$. It has negative curvature.

Problem

What curves form the other normal sections of the saddle above? Which ones have curvature 0?

Gauss's *Theorema Egregium*, to which we referred at the end of Chapter 5, can now be restated as *the product of the curvatures κ_1 and κ_2 can be determined from the intrinsic geometry of the surface.* Imagine that a piece of surface is made of material that is flexible but not stretchable. By flexing the material we can change the values of κ_1 and κ_2, but not their product. For example, take a sheet of typing paper. When it lies flat on the table, its curvatures κ_1 and κ_2 are both 0. If we curl the paper into a cylindrical tube, we can change κ_1 to some nonzero quantity, while leaving κ_2 alone. But we *cannot* curl the paper simultaneously in two different directions, for then the product $\kappa_1\kappa_2$ would no longer be zero. Experimentation with paper quickly leads one to this conclusion.

Another important example: Take a piece of a sphere. It is possible to pinch it so that its curvature increases in one direction and decreases in a perpendicular direction. This is consistent with the Theorema Egregium, which promises only that the product must remain constant. However, it is impossible to flatten out a piece of a sphere onto a plane without distortion. This means that no map of a portion of the Earth's surface can precisely scale all lengths and angles correctly. The Mercator map, for example, correctly displays angles (it is *conformal*) and takes longitudes and latitudes to straight lines. But it does not represent areas correctly: Greenland comes out looking too large, for instance.

While the product of the curvatures κ_1 and κ_2 is an intrinsic property of a surface, the average of the curvatures, $H = \frac{1}{2}(\kappa_1 + \kappa_2)$, is an extrinsic quantity, known as the *mean curvature*. One of the most important and well-studied problems in geometry concerns the question, What surfaces have the property that $H = 0$ everywhere? This question is important for the following reason: Suppose you take a loop of wire and dip it into a solution of bubble soap. When you remove it there may be a film of soap stretched around in a surface bounded by the wire loop. Surface tension causes the soap to contract to a surface that has the smallest possible area. Such a surface is called a *minimal surface*. Joseph Louis Lagrange formulated the fundamental principle of such surfaces in 1760: A minimal surface has $H = 0$ everywhere. This equation can be understood physically by use of Laplace's equation: The pressure difference p between the sides of a soap film at any point is proportional to the mean curvature H. So in order to be in equilibrium, the pressure has to be equalized, which means that the mean curvature has to be 0.

A soap bubble does not have mean curvature zero. Unlike a soap film, it encloses a volume of air, which resists compression. It turns out that soap bubbles satisfy the equation $H = C$, where C is some constant. Such surfaces are called surfaces of constant mean curvature. A sphere of radius $\frac{1}{H}$ satisfies this equation, and indeed, soap bubbles tend to be spherical. However, more complicated configurations of soap bubbles, in which several bubbles are stuck together, do occur. (See chapter 5 of [20] for a discussion of bubbles.)

If P is a point on a surface and $r > 0$ a small positive number, then the *geodesic circle of radius* r consists of all points Q for which the shortest path from P to Q has length r. In the plane, the circumference of such a circle is $2\pi r$. What does a geodesic circle of radius r look like on a sphere of radius R? For convenience, let P be the north pole and think of the sphere as the sphere of radius R centered at the origin O. Shortest paths from P are arcs of great circles of radius R centered at O and passing through the poles. If $\angle POQ$ is θ measured in radians, then the length of the arc from P to Q is $R\theta$. This will be r if

$$\theta = \frac{r}{R}.$$

Now we can see that a geodesic circle of radius r is a parallel of latitude lying in the plane $z = R - R\cos\frac{r}{R}$. This is a circle of radius $R\sin\frac{r}{R}$ in that plane, so it has circumference $2\pi R\sin\frac{r}{R}$. It can be shown using trigonometry that this quantity is smaller than $2\pi r$. The following fact can be established using elementary calculus:

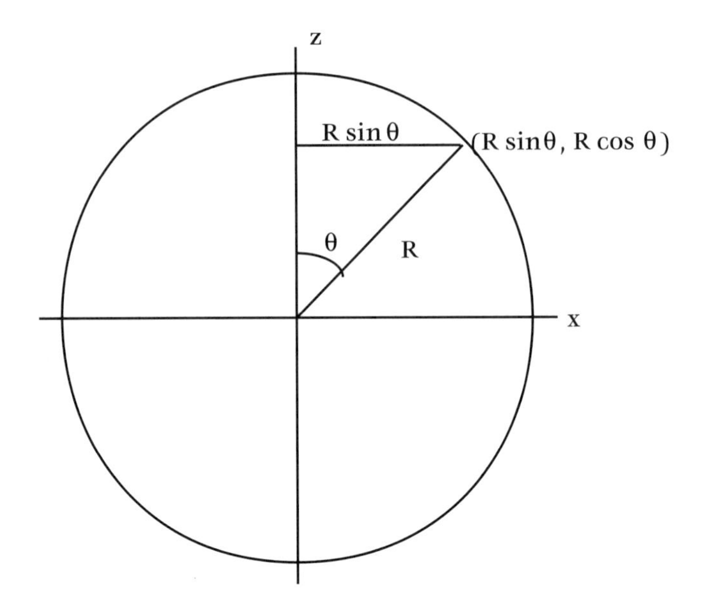

Proposition 6.2.1
Let $K = \frac{1}{R^2}$. Then on the sphere of radius R a geodesic circle of radius r has circumference $2\pi r - \frac{\pi}{3}Kr^3 + o(r^3)$, where $o(r^3)$ denotes a quantity that is very small compared with r^3.

Problem
Using a calculator or computer, compute the value of $2\pi R \sin \frac{r}{R}$ for r a small number and various values of R. Compare your answers with $2\pi r - \frac{\pi}{3}Kr^3$.

This proposition gives a new interpretation of Gaussian curvature. The sphere has curvature $K > 0$. This is reflected in having circles of radius r having circumference smaller than expected in Euclidean geometry. Straight lines bend toward each other, eventually even meeting. In a space with negative curvature, the opposite phenomenon occurs. The circumference of a circle turns out to be larger than the Euclidean value, as rays emanating from a point spread apart.

6.3 From Euclid to Einstein

What is the nature of the geometry of space? For over two thousand years the answer seemed to be contained in the work codified by Euclid. Until the nineteenth century it was believed that the laws of geometry were those that followed inevitably from his axioms and postulates.

In ancient Greek times, Aristotle formulated the theory that force was necessary to maintain motion; he observed that objects on Earth tended to move in straight lines when not influenced by outside forces. Galileo (1564–1642) developed the theory of motion, later restated by Newton in 1713 as his first law of motion: "Every body continues in its state of rest or of uniform motion in a straight line, except in so far as it may be compelled by force to change that state." This means that straight lines are a fundamental feature of space: They are the inertial paths of objects.

So how do these straight lines work? Until the birth of non-Euclidean geometry, it seemed that we had no choice. The laws of geometry were those of Euclid's geometry, and that was that. The discovery of hyperbolic geometry, and later of elliptic geometry, opened up new possibilities. But which one is right? and how do we decide.

It might seem that the way to answer the question would be to build a triangle, measure the angle sum, and compare with 180°. The problem is that we would need a really large triangle. I mean REALLY large. Remember that the defect (or excess) of a triangle, if it exists, is supposed to be proportional to its area. So the discrepancy between Euclidean and non-Euclidean triangles only becomes apparent when the size is quite large. Gauss is said to have attempted a measurement of a triangle formed by the peaks of three mountains, but the discrepancy he found was well within the range of experimental error.

As a practical matter, then, we may assume that the rules of Euclidean geometry hold in our own tiny corner of the galaxy. But Euclidean lines go on forever, while straight lines in elliptic geometry come back to themselves. This raises the intriguing question: If we assume that a light ray travels along a straight path, can I point a flashlight (albeit a ridiculously powerful one!) off into space and illuminate the back of my head?

Here is a chain of thoughts about how we might find out the answer to this question. Think of our universe as a three-dimensional geometric space. While we don't know what far distant portions of space look like, we will assume that they are geometrically about the same as our portion of space. This is the principle of *homogeneity*. In particular, this means that the "curvature" of space is the same everywhere. How is curvature defined in three dimensions? Riemann proposed a definition. Imagine at a point P a small piece of a straight line α. Draw the straight lines emanating from P in each direction perpendicular to α. If we are in Euclidean space, this family of lines fills out the plane perpendicular to α. But in general, this family will form a surface passing through the point P. Its curvature at P can be measured intrinsically (by Gauss's Theorema Egregium). Call this curvature $\kappa(\alpha)$. Do this for all possible straight lines through P; the resulting numbers describe the way space bends at P. The assumption of homogeneity is that these numbers will be the same at different points in space.

Next, assume that space is *isotropic*. This means that space looks the same in every direction; there are no "preferred" directions. A consequence of this is that the curvatures $\kappa(\alpha)$ would all be the same. Thus there is one number κ that describes how space bends in every direction at every point. This assumption is the assumption that the universe has *constant curvature*. It is then possible to prove mathematically that the universe is a three-dimensional version of one of our three geometries—Euclidean, hyperbolic, or elliptic—depending on whether κ is zero, negative, or positive.

Of course, we have been making rather generous assumptions. Why should space be homogeneous and isotropic? In a general Riemannian manifold the curvature can vary as we move from one point to another. On the surface of a (two-dimensional) torus, for example, the Gaussian curvature is positive on the outer part and negative on the inner part. To see this, note that the outer part has a convex shape, while the inner part has the shape of a saddle. This implies that if we construct a small equilateral triangle with sides of length ϵ in different parts of a torus (with geodesic sides), the angle sum will change. In a geometry with variable curvature, congruence of figures becomes a problem.

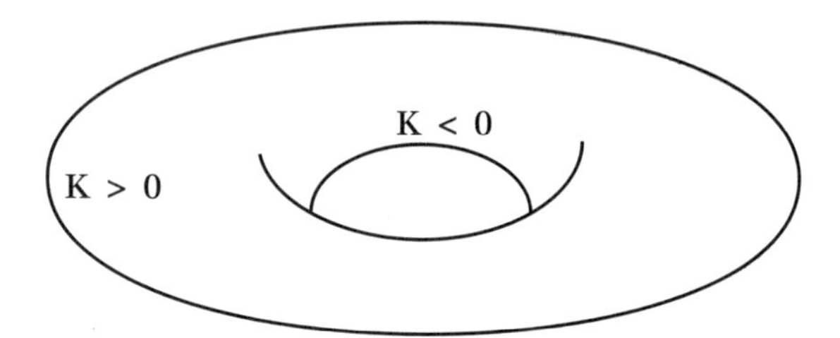

This is an argument for homogeneity. If physical objects can be moved from one place to another without changing their dimensions, then that would seem to suggest that space has to be homogeneous. If objects can be rotated without changing their properties, then space must be isotropic.

While appealing, this is not a convincing argument. For how do we know that objects do not change their shape as they move? Measuring them with rulers to see that their dimensions do not change only works if the rulers themselves do not change! In fact, if we assume only that space is approximately homogeneous and isotropic, then slight changes in shapes may be undetectable anyway.

The late nineteenth century saw another great change in our way of thinking about geometry. When James Clerk Maxwell developed his theory of electricity and magnetism, which appeared in 1871, he predicted that electromagnetic waves should exist, which would travel at a speed c, the speed of light. How is this speed to be measured? Presumably, with respect to the *ether*, the material of absolute space. (Just as sound propagates through air, so light was presumed to propagate through the ether.) In 1887, Albert A. Michelson and Edward W. Morley, working at institutions that later merged to form my school, Case Western Reserve University, devised and carried out an experiment to detect the ether and its effect on the speed of light. They reasoned that light travelling in different directions through the ether with respect to the motion of the earth would be observed to travel at different speeds. Their equipment was designed to be able to detect this difference between "upstream" and "downstream" motion.

The experiment showed no difference in the observed speed of light. This negative result led H.A. Lorentz to formulate a theory of "contraction," in which a moving object shrinks in the longitutidinal direction. In other words, a ruler moving along the path of a light ray is shorter than one staying in a fixed position. (Maybe that idea of objects changing shape as they move from place to place was not so fanciful after all.)

In 1905, Albert Einstein presented a new theory of space, the *special theory of relativity*, based on two assumptions: (1) The speed of light is a constant independent of the motion of the observer, and (2) all physical laws should have the same form for two observers moving at constant velocity relative to each other. His theory required that space and time be thought of as making up a single, *four-dimensional* entity known as space–time. The geometry of this space is not Euclidean but *Lorentzian*; its properties are quite different from those of Euclidean geometry, notably in the fact that it is no longer possible to speak about "lengths."

Let us look briefly at the two-dimensional version of Lorentzian geometry. Our model is the plane $\mathcal{M}$, which we imagine as having two coordinate axes x and t. The x coordinate stands for the position of an object in space and the t coordinate for the position in time. An object moving at constant speed v will go from a position x at time t to a position $x + vh$ at time $t + h$. As h varies, the object traces out a path (called the "world line") that is a straight line.

If $v = 0$ (an object standing still), the world line is a vertical line $x = k$; the faster the object moves, the less steep the slope of the line. If we assume that no object travels faster than c, the speed of light, and if we choose units of measurement such that $c = 1$, then no world line can tilt at an angle less than 45° from the horizontal. On the other hand, light rays travel along straight lines that have exactly 45° slope.

Our choice of coordinate axes gives us a reference point (the origin $(0, 0)$) corresponding to designating a point in space as a reference and a moment in time as a reference. If we change to a new reference point $Q = (c, d)$, nothing should change in our description of the geometry of

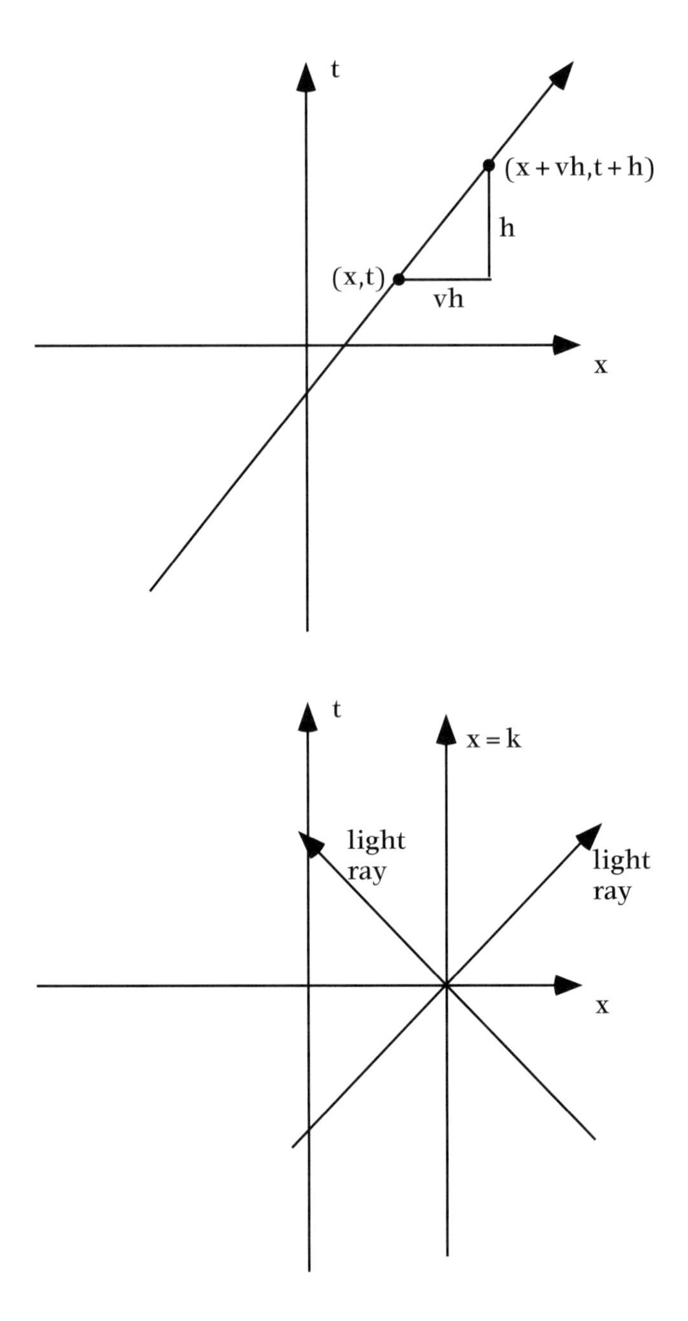

space–time. This means that we can pick a new set of coordinate axes parallel to the old one with the origin located at Q. Geometrically, we allow all (Euclidean) translations $\mathbf{T}$ to be considered isometries of our space $\mathcal{M}$. The new coordinate system (x', t') is related to the old one by the formulas

$$x' = x - c, \quad t' = t - d.$$

Newton's law of motion, which says that the acceleration of an object is proportional to the (vector) sum of the forces acting on it, implies the existence of a coordinate system with respect to which the law holds. Such a coordinate system is called an *inertial system*. In other coordinate systems the law will not hold; for example, in a rotating coordinate system, objects do not follow the paths that would be predicted by assuming an inertial coordinate system. Newtonian relativity is the principle that says that the law of motion will remain unchanged in form with respect to another coordinate system moving relative to an inertial system at constant velocity. In a Newtonian space–time, we would therefore allow a new coordinate system (x', t') given by $x' = x - vt$, $t' = t$. In other words, a point P that has coordinates (a, b) with respect to the x-t system will have new coordinates $(a - vb, b)$. The new time-axis is tilted with respect to the old one, while the space-axis does not change. (It may seem that the x-axis should be the one to change; study the picture below to see why that does not happen.)

In the new coordinate system, an object that had been traveling at a constant speed w would be observed to move at the constant speed $w - v$. For instance, any object moving at speed v (with respect to our old coordinate system) now appears to be standing still. This principle

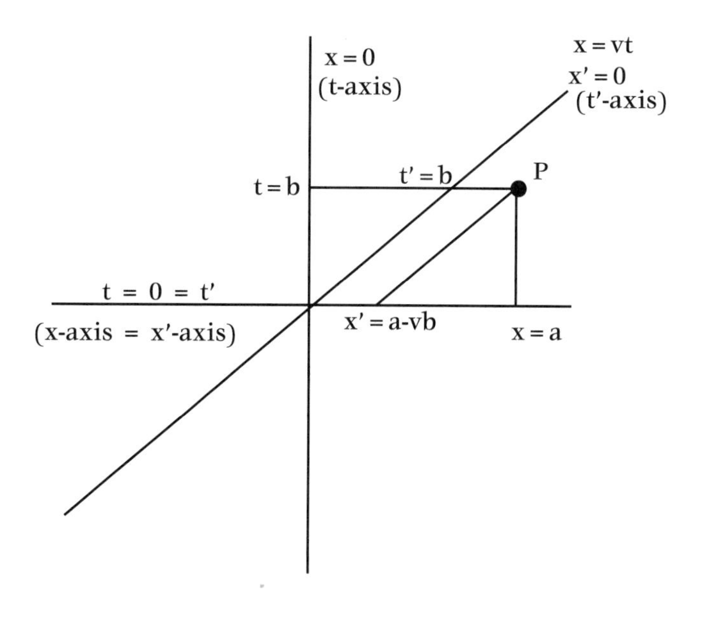

would hold even for light rays. But the Michelson–Morley experiment showed that this is not the case. Light rays are always observed to move at the speed $c = 1$. The coordinate change described above is therefore not allowed in Einstein's theory of relativity. Instead, it turns out that the appropriate coordinate change is given by the formulas

$$x' = \frac{x - vt}{\sqrt{1 - v^2}}, \quad t' = \frac{t - vx}{\sqrt{1 - v^2}}. \tag{6.3.1}$$

Problem

Verify that this transformation carries the equation $x = t$ to the equation $x' = t'$, so the speed of a light ray remains unchanged. More generally, If an object moves at velocity u, so that $x = ut + b$, with $-1 \leq u \leq 1$, what is the velocity measured in the (x', t') coordinate system?

The transformations described above are called *Lorentz transformations*; they form a group under composition. The space $\mathcal{M}$ is called (2-dimensional) *Minkowski space*, named after the Russian mathematician Hermann Minkowski. A geometric object in this space should be something whose description remains unchanged under Lorentz transformations. The property of being a straight line is one example. There are three types of straight lines: *timelike*, which can be made the t–axis in some coordinate system, *spacelike*, which can be made the x-axis, and *light–like*, which trace out the paths of light rays.

Circles are not generally taken to circles, but there is another class of curves that play a role similar to that of circles in Euclidean geometry. The curve C given by the equation $x^2 - t^2 = c$, $c \neq 0$, is preserved under all Lorentz transformations. In fact, given two points A and B on the curve C, there is exactly one transformation of the form 6.3.1 that takes A to B. We may think of curves like C as "circles" centered at the origin and the Lorentz transformations as "rotations" of these circles. (Actually, "translations" is probably a more appropriate term.)

Problem

Verify the assertions made in the last paragraph.

Our discussion has concerned two dimensions. Three–dimensional Minkowski space $\mathcal{M}^3$ has two space coordinates. If we spin our two-dimensional model around the t–axis, we get the three-dimensional version. Instead of light traveling in two possible directions, we have a circle's worth of directions. The possible paths taken by a light ray passing through a point P in $\mathcal{M}^3$ form a *cone*, called the *light cone*. The curves described above are replaced by surfaces $x^2 + y^2 - t^2 = c$. When $c = 0$, this is an equation of the light cone through the origin. When $c > 0$, the

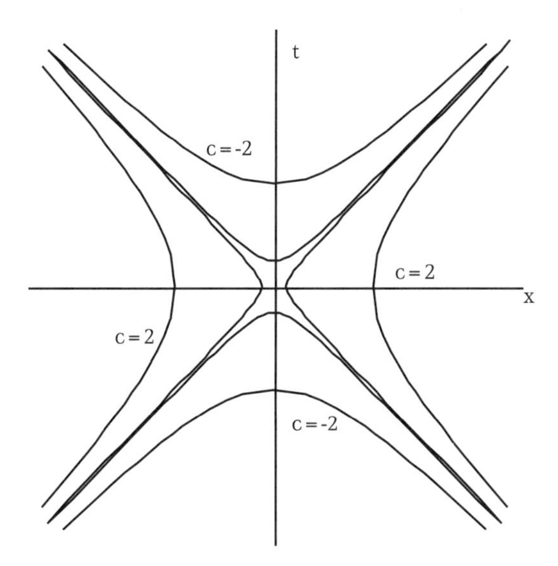

surface is the *hyperboloid of one sheet*. When $c < 0$, the surface has two pieces: It is the *hyperboloid of two sheets*. The Lorentz transformations fix the origin, take the light cone to itself, and take each hyperboloid to itself. Rotations about the t axis also preserve the light cone and the hyperboloids.

We may think of the surface $t^2 - x^2 - y^2 = 1$, $t > 0$, as a geometric surface and the rotations and Lorentz transformations as isometries of that surface. The remarkable fact is that this surface is then a model of the hyperbolic plane! Likewise, in four-dimensional space–time, the space $t^2 - x^2 - y^2 - z^2 = 1$ is a model for three-dimensional hyperbolic geometry. Thus hyperbolic geometry turns out to play a significant role in the geometry of space–time.

The quantity $t^2 - x^2 - y^2 - z^2$ is called the *Minkowskian distance* from the origin to the point with coordinates (x, y, z, t). This is a little bit like Euclidean distance, which would be $\sqrt{x^2 + y^2 + z^2 + t^2}$. Notice that the Minkowskian distance can be negative or zero, so it would not be a good idea to take a square root. For a physical interpretation of this distance, as well as a discussion of clocks, black holes, and other interesting phenomena, you may want to read the expository article by Roger Penrose "Geometry of the Universe," in [33].

In 1916, Einstein generalized his theory to include the role of gravity in influencing motion. The *general theory of relativity* describes the universe as a four-dimensional space–time, in which gravitational fields are manifested through curvature.

This idea can be illustrated intuitively. Suppose, for example, we want to understand how a satellite orbits around the earth in a spiraling path ending in the satellite hitting the ground. Picture a taut bedsheet with a heavy object placed in the middle of it. The weight causes the sheet to sag in the middle, forming a well. A ball rolling around the sag will spiral inward, eventually dropping into the center.

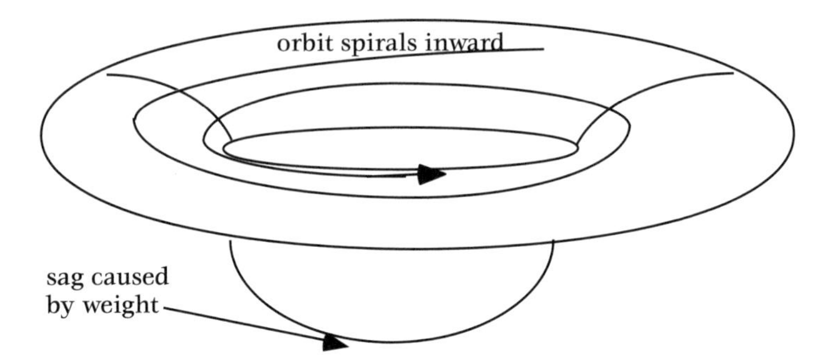

Just as Riemann formulated a general idea of a space in which the geometry was "locally Euclidean," while possibly having variable shape as we move from one place to another, so Einstein proposed a notion of a space as being "locally Lorentzian," with the presence of gravitational objects being reflected in changes in the shape of space from one place to another. The mathematical study of Riemann's spaces is called Riemannian geometry, while Einstein's space falls under a more general heading called pseudo-Riemannian geometry.

This grandly complicated description of the universe seems far distant from the geometry described by Euclid 2,300 years ago. And yet, had it not been for Euclid's formulation of the parallel postulate, we might never have arrived at this point in our understanding of space. By insisting on making the property of parallel lines a Postulate, he initiated a process that led, 2,100 years later, to the discovery that he was right in doing so. That gave rise to the thought that there was more than one way to think about the mathematical principles of geometry, which led to the great leaps of the nineteenth and twentieth centuries.

References

[1] Simon L. Altmann, *Rotations, Quaternions, and Double Groups*, Oxford University Press, Oxford, 1986.

[2] W.W. Rouse Ball and H.S.M. Coxeter, *Mathematical Recreations and Essays*, 13th edition, Dover Publications, New York, 1987.

[3] David Barnette, *Map Coloring, Polyhedra, and the Four-Color Problem*, Dolciani Mathematical Expositions No. 8, Mathematical Association of America, 1983.

[4] Roberto Bonola, *Non-Euclidean Geometry*, Dover Publications, New York, 1955.

[5] J. Bourgoin, *Arabic Geometrical Pattern and Design*, Dover Publications, New York, 1973.

[6] Robert Connelly, "Rigidity", in *Handbook of Convex Geometry*, ed. by P.M. Gruber and J.M. Wills, Elsevier Science Publishers B.V., 1993.

[7] H.S.M. Coxeter, *Introduction to Geometry*, John Wiley, New York, 1963.

[8] H.S.M. Coxeter, *Regular Polytopes*, Dover Publications, New York, 1963.

[9] Howard Eves, *Great Moments in Mathematics After 1650*, Dolciani Mathematical Expositions **7**, MAA, 1983.

[10] Howard Eves, *An Introduction to the Foundations and Fundamental Concepts of Mathematics*, Rinehart and Company, Inc., New York, 1958.

[11] Howard Eves, *An Introduction to the History of Mathematics*, Holt, Rinehart and Winston, New York, 1953.

[12] K.J. Falconer, *the Geometry of Fractal Sets*, Cambridge University Press, Cambridge, 1985.

[13] Martin Gardner, *The New Ambidextrous Universe*, 3rd ed., W.H. Freeman and Company, New York, 1990.

[14] Martin Gardner, *Penrose Tiles to Trapdoor Ciphers*, W.H. Freeman and Company, New York, 1989.

[15] Martin Gardner, *The New Ambidextrous Universe*, W.H. Freeman and Company, New York, 1990.

[16] Martin Gardner, *New Mathematical Diversions from Scientific American*, Simon and Schuster, New York, 1966.

[17] Martin Gardner, *Time Travel and Other Mathematical Bewilderments*, W.H. Freeman and Company, New York, 1988.

[18] T. L. Heath, *The Thirteen Books of Euclid's Elements, vol. 1*,Dover Publications, New York, 1956.

[19] T. L. Heath, *The Thirteen Books of Euclid's Elements, vol. 3*,Dover Publications, New York, 1956.

[20] Stefan Hildebrandt and Anthony Tromba, *Mathematics and Optimal Form*, Scientific American Books, Inc., New York, 1985.

[21] David Hilbert, *Foundations of Geometry*, tenth German edition, English translation by Leo Unger, Open Court Publishing Company, La Salle, Illinois, 1994.

[22] Edward Kasner and James Newman, *Mathematics and the Imagination*, Simon and Schuster, New York, 1940.

[23] Maurice Kraitchik, *Mathematical Recreations*, Dover Publications, New York, 1953.

[24] L.A. Lyusternik, *Convex Figures and Polyhedra*, D.C. Heath and Company, Boston, 1966.

[25] Benoit Mandelbrot, *The Fractal Geometry of Nature*, W.H. Freeman and Company, New York, 1982.

[26] George E. Martin, *Transformation Geometry*, Springer-Verlag, New York, 1982.

[27] Moritz Pasch, *Vorlesungen über neuere Geometrie*, B.G. Teubner, Leipzig, 1882.

[28] Dan Pedoe, *Circles, a Mathematical View*, Dover Publications, New York, 1979.

[29] Dan Pedoe, *Geometry, a Comprehensive Course*, Dover Publications, New York, 1988.

[30] Doris Schattschneider, *Visions of Symmetry*, W.H. Freeman and Company, New York, 1990.

[31] Marjorie Senechal and George Fleck, *Shaping Space: a Polyhedral Approach*, Birkhäuser Boston, Inc., 1988.

[32] Michael Spivak, *A Comprehensive Introduction to Differential Geometry*, five volumes, Publish or Perish, Inc., Boston, 1975.

[33] Lynn Steen, *Mathematics Today*, Vintage Books, New York, 1980.

[34] John Stillwell, *Classical Topology and Combinatorial Group Theory*, Springer-Verlag, New York, 1980.

[35] Richard J. Trudeau, *The Non-Euclidean Revolution*, Birkhäuser Boston, 1987.

[36] Harold E. Wolfe *Introduction to Non-Euclidean Geometry*, Holt, Rinehart and Winston, New York, 1945.

Index

(continued from page ii)

Troutman: Variational Calculus and Optimal Control. Second edition.
Valenza: Linear Algebra: An Introduction to Abstract Mathematics.
Whyburn/Duda: Dynamic Topology.
Wilson: Much Ado About Calculus.